Teacher Edition

Eureka Math
Grade 1
Module 6

Special thanks go to the Gordon A. Cain Center and to the Department of Mathematics at Louisiana State University for their support in the development of *Eureka Math*.

Published by the non-profit Great Minds

Copyright © 2015 Great Minds. No part of this work may be reproduced, sold, or commercialized, in whole or in part, without written permission from Great Minds. Non-commercial use is licensed pursuant to a Creative Commons Attribution-NonCommercial-ShareAlike 4.0 license; for more information, go to http://greatminds.net/maps/math/copyright. "Great Minds" and "Eureka Math" are registered trademarks of Great Minds.

Printed in the U.S.A.
This book may be purchased from the publisher at eureka-math.org
10 9 8 7 6 5 4 3 2

ISBN 978-1-63255-353-9

Eureka Math: A Story of Units Contributors

Katrina Abdussalaam, Curriculum Writer
Tiah Alphonso, Program Manager—Curriculum Production
Kelly Alsup, Lead Writer / Editor, Grade 4
Catriona Anderson, Program Manager—Implementation Support
Debbie Andorka-Aceves, Curriculum Writer
Eric Angel, Curriculum Writer
Leslie Arceneaux, Lead Writer / Editor, Grade 5
Kate McGill Austin, Lead Writer / Editor, Grades PreK–K
Adam Baker, Lead Writer / Editor, Grade 5
Scott Baldridge, Lead Mathematician and Lead Curriculum Writer
Beth Barnes, Curriculum Writer
Bonnie Bergstresser, Math Auditor
Bill Davidson, Fluency Specialist
Jill Diniz, Program Director
Nancy Diorio, Curriculum Writer
Nancy Doorey, Assessment Advisor
Lacy Endo-Peery, Lead Writer / Editor, Grades PreK–K
Ana Estela, Curriculum Writer
Lessa Faltermann, Math Auditor
Janice Fan, Curriculum Writer
Ellen Fort, Math Auditor
Peggy Golden, Curriculum Writer
Maria Gomes, Pre-Kindergarten Practitioner
Pam Goodner, Curriculum Writer
Greg Gorman, Curriculum Writer
Melanie Gutierrez, Curriculum Writer
Bob Hollister, Math Auditor
Kelley Isinger, Curriculum Writer
Nuhad Jamal, Curriculum Writer
Mary Jones, Lead Writer / Editor, Grade 4
Halle Kananak, Curriculum Writer
Susan Lee, Lead Writer / Editor, Grade 3
Jennifer Loftin, Program Manager—Professional Development
Soo Jin Lu, Curriculum Writer
Nell McAnelly, Project Director

Ben McCarty, Lead Mathematician / Editor, PreK–5
Stacie McClintock, Document Production Manager
Cristina Metcalf, Lead Writer / Editor, Grade 3
Susan Midlarsky, Curriculum Writer
Pat Mohr, Curriculum Writer
Sarah Oyler, Document Coordinator
Victoria Peacock, Curriculum Writer
Jenny Petrosino, Curriculum Writer
Terrie Poehl, Math Auditor
Robin Ramos, Lead Curriculum Writer / Editor, PreK–5
Kristen Riedel, Math Audit Team Lead
Cecilia Rudzitis, Curriculum Writer
Tricia Salerno, Curriculum Writer
Chris Sarlo, Curriculum Writer
Ann Rose Sentoro, Curriculum Writer
Colleen Sheeron, Lead Writer / Editor, Grade 2
Gail Smith, Curriculum Writer
Shelley Snow, Curriculum Writer
Robyn Sorenson, Math Auditor
Kelly Spinks, Curriculum Writer
Marianne Strayton, Lead Writer / Editor, Grade 1
Theresa Streeter, Math Auditor
Lily Talcott, Curriculum Writer
Kevin Tougher, Curriculum Writer
Saffron VanGalder, Lead Writer / Editor, Grade 3
Lisa Watts-Lawton, Lead Writer / Editor, Grade 2
Erin Wheeler, Curriculum Writer
MaryJo Wieland, Curriculum Writer
Allison Witcraft, Math Auditor
Jessa Woods, Curriculum Writer
Hae Jung Yang, Lead Writer / Editor, Grade 1

Board of Trustees

Lynne Munson, President and Executive Director of Great Minds
Nell McAnelly, Chairman, Co-Director Emeritus of the Gordon A. Cain Center for STEM Literacy at Louisiana State University
William Kelly, Treasurer, Co-Founder and CEO at ReelDx
Jason Griffiths, Secretary, Director of Programs at the National Academy of Advanced Teacher Education
Pascal Forgione, Former Executive Director of the Center on K-12 Assessment and Performance Management at ETS
Lorraine Griffith, Title I Reading Specialist at West Buncombe Elementary School in Asheville, North Carolina
Bill Honig, President of the Consortium on Reading Excellence (CORE)
Richard Kessler, Executive Dean of Mannes College the New School for Music
Chi Kim, Former Superintendent, Ross School District
Karen LeFever, Executive Vice President and Chief Development Officer at ChanceLight Behavioral Health and Education
Maria Neira, Former Vice President, New York State United Teachers

A STORY OF UNITS

1 GRADE

Mathematics Curriculum

GRADE 1 • MODULE 6

Table of Contents

GRADE 1 • MODULE 6

Place Value, Comparison, Addition and Subtraction to 100

Module Overview 2

Topic A: Comparison Word Problems 10

Topic B: Numbers to 120.. 41

Topic C: Addition to 100 Using Place Value Understanding................................. 132

Topic D: Varied Place Value Strategies for Addition to 100 214

Mid-Module Assessment and Rubric 235

Topic E: Coins and Their Values.. 247

Topic F: Varied Problem Types Within 20 302

End-of-Module Assessment and Rubric 341

Topic G: Culminating Experiences 353

Answer Key 375

Module 1: Place Value, Comparison, Addition and Subtraction to 100

1

Grade 1 • Module 6
Place Value, Comparison, Addition and Subtraction to 100

OVERVIEW

In this final module of the Grade 1 curriculum, students bring together their learning from Module 1 through Module 5 to learn the most challenging Grade 1 standards and celebrate their progress.

In Topic A, students grapple with comparative word problem types (**1.OA.1**). While students solved some comparative problem types during Module 3 and within the Application Problems in Module 5, this is their first opportunity to name these types of problems and learn to represent comparisons using tape diagrams with two tapes.

Students extend their understanding of and skill with tens and ones to numbers to 100 in Topic B (**1.NBT.2**). For example, they mentally find 10 more, 10 less, 1 more, and 1 less (**1.NBT.5**) and compare numbers using the symbols >, =, and < (**1.NBT.3**). They then count and write numbers to 120 (**1.NBT.1**) using both standard numerals and the unit form.

In Topics C and D, students again extend their learning from Module 4 to the numbers to 100 to add and subtract (**1.NBT.4**, **1.NBT.6**). They add pairs of two-digit numbers in which the ones digits sometimes have a sum greater than 10, recording their work using various methods based on place value (**1.NBT.4**). In Topic D, students focus on using drawings, numbers, and words to solve, highlighting the role of place value, the properties of addition, and related facts.

At the start of the second half of Module 6, students are introduced to nickels and quarters (**1.MD.3**), having already used pennies and dimes in the context of their work with numbers to 40 in Module 4. Students use their knowledge of tens and ones to explore decompositions of the values of coins. For example, they might represent 25 cents using 1 quarter, 25 pennies, 2 dimes and 1 nickel, or 1 dime and 15 pennies.

In Topic F, students really dig into MP.1 and MP.3. The topic includes the more challenging *compare with bigger or smaller unknown* word problem types, wherein *more* or *less* suggests the incorrect operation (**1.OA.1**), thus giving a context for more in-depth discussions and critiques. On the final day of this topic, students work with varied problem types, sharing and explaining their strategies and reasoning. Peers ask each other questions and defend their choices. The End-of-Module Assessment follows Topic F.

The module and year close with Topic G, wherein students celebrate their year's worth of learning with fun fluency festivities that equip them with games to maintain their fluency during the summer months prior to Grade 2. To send home their year's work, the final day is devoted to creating a math folder illustrating their learning.

Module Overview 1•6

Notes on Pacing for Differentiation

During Module 4, addition and subtraction work is limited to numbers within 40. In Module 6, students extend into numbers within 100. If students are readily able to apply their learning from Module 4 to Module 6, consider consolidating lessons in Topics A, B, and C (e.g., Lessons 3 and 4, Lessons 5 and 6, and Lessons 10 and 11). In Topic C, use each day's Exit Ticket to determine whether the lessons that follow can be omitted or consolidated.

Topic E, Coins and Their Values, might be modified, omitted, or embedded throughout the instructional day depending on the standards in the state implementing the curriculum.

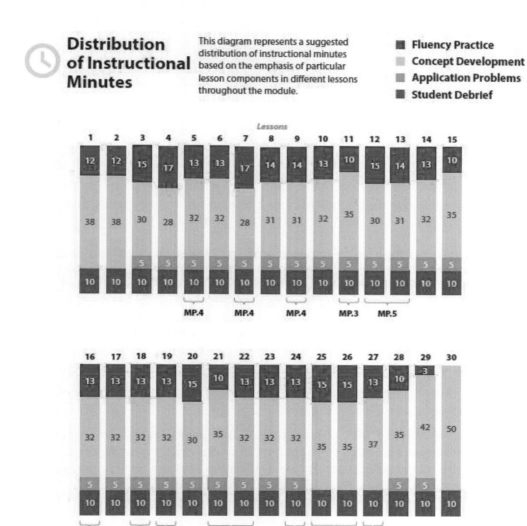

Module 1: Place Value, Comparison, Addition and Subtraction to 100

Focus Grade Level Standards

Represent and solve problems involving addition and subtraction.

1.OA.1   Use addition and subtraction within 20 to solve word problems involving situations of adding to, taking from, putting together, taking apart, and comparing, with unknowns in all positions, e.g., by using objects, drawings, and equations with a symbol for the unknown number to represent the problem. (See CCSS-M Glossary, Table 1.)

Extend the counting sequence.

1.NBT.1   Count to 120, starting at any number less than 120. In this range, read and write numerals and represent a number of objects with a written numeral.

Understand place value.

1.NBT.2   Understand that the two digits of a two-digit number represent amounts of tens and ones. Understand the following as special cases:

   a.   10 can be thought of as a bundle of ten ones—called a "ten."

   c.   The numbers 10, 20, 30, 40, 50, 60, 70, 80, 90 refer to one, two, three, four, five, six, seven, eight, or nine tens (and 0 ones).

1.NBT.3   Compare two two-digit numbers based on meanings of the tens and ones digits, recording the results of comparisons with the symbols >, =, and <.

Use place value understanding and properties of operations to add and subtract.

1.NBT.4   Add within 100, including adding a two-digit number and a one-digit number, and adding a two-digit number and a multiple of 10, using concrete models or drawings and strategies based on place value, properties of operations, and/or the relationship between addition and subtraction; relate the strategy to a written method and explain the reasoning used. Understand that in adding two-digit numbers, one adds tens and tens, ones and ones; and sometimes it is necessary to compose a ten.

1.NBT.5   Given a two-digit number, mentally find 10 more or 10 less than the number, without having to count; explain the reasoning used.

1.NBT.6   Subtract multiples of 10 in the range 10–90 from multiples of 10 in the range 10–90 (positive or zero differences), using concrete models or drawings and strategies based on place value, properties of operations, and/or the relationship between addition and subtraction; relate the strategy to a written method and explain the reasoning used.

Tell and write time and money.[1]

1.MD.3   Tell and write time in hours and half-hours using analog and digital clocks. Recognize and identify coins, their names, and their value.

Foundational Standards

K.OA.2   Solve addition and subtraction word problems, and add and subtract within 10, e.g., by using objects or drawings to represent the problem.

K.OA.3   Decompose numbers less than or equal to 10 into pairs in more than one way, e.g., by using objects or drawings, and record each decomposition by a drawing or equation (e.g., 5 = 2 + 3 and 5 = 4 + 1).

K.OA.4   For any number from 1 to 9, find the number that makes 10 when added to the given number, e.g., by using objects or drawings, and record the answer with a drawing or equation.

K.NBT.1   Compose and decompose numbers from 11 to 19 into ten ones and some further ones, e.g., by using objects or drawings, and record each composition or decomposition by a drawing or equation (e.g., 18 = 10 + 8); understand that these numbers are composed of ten ones and one, two, three, four, five, six, seven, eight, or nine ones.

Focus Standards for Mathematical Practice

MP.1   **Make sense of problems and persevere in solving them.** Throughout Topic A, students analyze given situations and determine whether they are compare, take away, or put together problem types. Students' drawings, such as single and double tape diagrams, represent their planning toward a solution pathway. During Topic F, students initially work independently, supporting them in learning how to persevere and make sense of problems. As students share their strategies and solutions asking and answering peer questions, they demonstrate understanding of the approaches of their peers and identify corresponding elements between the approaches.

MP.3   **Construct viable arguments and critique the reasoning of others.** During Topic F, students share their strategies and reasoning as they explain their solutions to various problem types. They ask useful questions to help clarify or improve peers' explanations, such as, "How does your drawing help demonstrate your thinking?" Students consider how a selected student's work helped her solve the problem, as well as other pathways for the student to correctly solve the problem. As students share their thinking, they explain the mathematical reasoning that supports their argument.

---

[1] Time is addressed in Module 5. This module addresses the second portion of the standard, regarding money, which was added by New York State. Check your state and local standards to determine whether money is an expectation for your students.

MP.4		**Model with mathematics.** Throughout this module, students model their mathematics in various ways. While problem solving, students use tape diagrams and number sentences to model situations and solutions. When sharing various strategies for adding within 100, students use number bonds, number sentences, and sometimes drawings to solve for the sums and to demonstrate their understanding and use of place value, properties of addition, and the relationship between addition and subtraction as they decompose and recompose numbers.	
MP.5		**Use appropriate tools strategically.** After learning varied representations and strategies for adding and subtracting pairs of two-digit numbers, students choose their preferred methods for representing and solving problems efficiently. As they share their strategies, students explain their choice of making ten, adding tens and then ones, or adding ones and then tens. They also demonstrate how their choice of written method (number bonds, vertical alignment, or arrow notation) expresses their strategy work.	

Overview of Module Topics and Lesson Objectives

Standards		Topics and Objectives		Days
1.OA.1	A	**Comparison Word Problems**		2
		Lesson 1:	Solve *compare with difference unknown* problem types.	
		Lesson 2:	Solve *compare with bigger or smaller unknown* problem types.	
1.NBT.1 1.NBT.2a 1.NBT.2c 1.NBT.3 1.NBT.5	B	**Numbers to 120**		7
		Lesson 3:	Use the place value chart to record and name tens and ones within a two-digit number up to 100.	
		Lesson 4:	Write and interpret two-digit numbers to 100 as addition sentences that combine tens and ones.	
		Lesson 5:	Identify 10 more, 10 less, 1 more, and 1 less than a two-digit number within 100.	
		Lesson 6:	Use the symbols >, =, and < to compare quantities and numerals to 100.	
		Lesson 7:	Count and write numbers to 120. Use Hide Zero cards to relate numbers 0 to 20 to 100 to 120.	
		Lesson 8:	Count to 120 in unit form using only tens and ones. Represent numbers to 120 as tens and ones on the place value chart.	
		Lesson 9:	Represent up to 120 objects with a written numeral.	
1.NBT.4 1.NBT.6	C	**Addition to 100 Using Place Value Understanding**		8
		Lesson 10:	Add and subtract multiples of 10 from multiples of 10 to 100, including dimes.	

Standards	Topics and Objectives		Days
	Lesson 11:	Add a multiple of 10 to any two-digit number within 100.	
	Lesson 12:	Add a pair of two-digit numbers when the ones digits have a sum less than or equal to 10.	
	Lessons 13–14:	Add a pair of two-digit numbers when the ones digits have a sum greater than 10 using decomposition.	
	Lesson 15:	Add a pair of two-digit numbers when the ones digits have a sum greater than 10 with drawing. Record the total below.	
	Lessons 16–17:	Add a pair of two-digit numbers when the ones digits have a sum greater than 10 with drawing. Record the new ten below.	
1.NBT.4	D	**Varied Place Value Strategies for Addition to 100**	2
	Lesson 18:	Add a pair of two-digit numbers with varied sums in the ones, and compare the results of different recording methods.	
	Lesson 19:	Solve and share strategies for adding two-digit numbers with varied sums.	
	Mid-Module Assessment: Topics A–D (assessment 1 day, return 1 day, remediation or further applications 1 day)		3
1.MD.3	E	**Coins and Their Values**	5
	Lesson 20:	Identify pennies, nickels, and dimes by their image, name, or value. Decompose the values of nickels and dimes using pennies and nickels.	
	Lesson 21:	Identify quarters by their image, name, or value. Decompose the value of a quarter using pennies, nickels, and dimes.	
	Lesson 22:	Identify varied coins by their image, name, or value. Add one cent to the value of any coin.	
	Lesson 23:	Count on using pennies from any single coin.	
	Lesson 24:	Use dimes and pennies as representations of numbers to 120.	
1.OA.1	F	**Varied Problem Types Within 20**	3
	Lessons 25–26:	Solve *compare with bigger or smaller unknown* problem types.	
	Lesson 27:	Share and critique peer strategies for solving problems of varied types.	
	End-of-Module Assessment: Topics A–F (assessment 1 day, return ½ day, remediation or further applications ½ day)		2

Module 1: Place Value, Comparison, Addition and Subtraction to 100

Standards	Topics and Objectives	Days
	G **Culminating Experiences** Lessons 28–29: Celebrate progress in fluency with adding and subtracting within 10 (and 20). Organize engaging summer practice. Lesson 30: Create folder covers for work to be taken home illustrating the year's learning.	3
Total Number of Instructional Days		**35**

Terminology

New or Recently Introduced Terms

- Dime
- Nickel
- Penny
- Quarter

Familiar Terms and Symbols[2]

- <, >, = (less than, greater than, equal to)

Suggested Tools and Representations

- 100-bead Rekenrek
- Tape diagram

Homework

Homework at the K–1 level is not a convention in all schools. In this curriculum, homework is an opportunity for additional practice of the content from the day's lesson. The teacher is encouraged, with the support of parents, administrators, and colleagues, to discern the appropriate use of homework for his or her students. Fluency exercises can also be considered as an alternative homework assignment.

---

[2] These are terms and symbols students have seen previously.

Scaffolds[3]

The scaffolds integrated into *A Story of Units* give alternatives for how students access information as well as express and demonstrate their learning. Strategically placed margin notes are provided within each lesson elaborating on the use of specific scaffolds at applicable times. They address many needs presented by English language learners, students with disabilities, students performing above grade level, and students performing below grade level. Many of the suggestions are organized by Universal Design for Learning (UDL) principles and are applicable to more than one population. To read more about the approach to differentiated instruction in *A Story of Units*, please refer to "How to Implement *A Story of Units*."

Assessment Summary

Type	Administered	Format	Standards Addressed
Mid-Module Assessment Task	After Topic D	Constructed response with rubric	1.OA.1 1.NBT.1 1.NBT.2a 1.NBT.2c 1.NBT.3 1.NBT.4 1.NBT.5 1.NBT.6
End-of-Module Assessment Task	After Topic F	Constructed response with rubric	1.OA.1 1.NBT.1 1.NBT.2a 1.NBT.2c 1.NBT.3 1.NBT.4 1.NBT.5 1.NBT.6 1.MD.3[4]

---

[3] Students with disabilities may require Braille, large print, audio, or special digital files. Please visit the website www.p12.nysed.gov/specialed/aim for specific information on how to obtain student materials that satisfy the National Instructional Materials Accessibility Standard (NIMAS) format.

[4] Focus on money.

A STORY OF UNITS

Mathematics Curriculum

GRADE 1 • MODULE 6

Topic A
Comparison Word Problems

1.OA.1

Focus Standard:	1.OA.1	Use addition and subtraction within 20 to solve word problems involving situations of adding to, taking from, putting together, taking apart, and comparing, with unknowns in all positions, e.g., by using objects, drawings, and equations with a symbol for the unknown number to represent the problem. (See CCSS-M Glossary, Table 1.)
Instructional Days:	2	
Coherence -Links from:	G1–M3	Ordering and Comparing Length Measurements as Numbers
	G1–M4	Place Value, Comparison, Addition and Subtraction to 40
-Links to:	G2–M7	Problem Solving with Length, Money, and Data

Topic A of Module 6 opens with students exploring one of the most challenging problem types for their grade level, comparison word problems (see Table 2 below from *Counting and Cardinality and Operations and Algebraic Thinking Progressions* document, page 9) (**1.OA.1**). Students were informally introduced to the problem type in Module 3 as they analyzed data and compared measurements. During Module 5, students worked with comparison contexts through Application Problems. It is with this background that teachers can make informed choices during Module 6 to support students in recognizing and solving comparison word problems.

In Lesson 1, students work with *compare with difference unknown* problem types using double tape diagrams. They then carry their understanding of double tape diagrams into Lesson 2 to tackle *compare with bigger or smaller unknown* problem types. Throughout the module, students continue to practice these problem types as they solve Application Problems in the topics that follow.

Topic A: Comparison Word Problems

Table 2: Addition and subtraction situations by grade level.

	Result Unknown	Change Unknown	Start Unknown
Add To	A bunnies sat on the grass. B more bunnies hopped there. How many bunnies are on the grass now? $A + B = \square$	A bunnies were sitting on the grass. Some more bunnies hopped there. Then there were C bunnies. How many bunnies hopped over to the first A bunnies? $A + \square = C$	Some bunnies were sitting on the grass. B more bunnies hopped there. Then there were C bunnies. How many bunnies were on the grass before? $\square + B = C$
Take From	C apples were on the table. I ate B apples. How many apples are on the table now? $C - B = \square$	C apples were on the table. I ate some apples. Then there were A apples. How many apples did I eat? $C - \square = A$	Some apples were on the table. I ate B apples. Then there were A apples. How many apples were on the table before? $\square - B = A$

	Total Unknown	Both Addends Unknown[1]	Addend Unknown[2]
Put Together /Take Apart	A red apples and B green apples are on the table. How many apples are on the table? $A + B = \square$	Grandma has C flowers. How many can she put in her red vase and how many in her blue vase? $C = \square + \square$	C apples are on the table. A are red and the rest are green. How many apples are green? $A + \square = C$ $C - A = \square$

	Difference Unknown	Bigger Unknown	Smaller Unknown
Compare	"How many more?" version. Lucy has A apples. Julie has C apples. How many more apples does Julie have than Lucy? "How many fewer?" version. Lucy has A apples. Julie has C apples. How many fewer apples does Lucy have than Julie? $A + \square = C$ $C - A = \square$	"More" version suggests operation. Julie has B more apples than Lucy. Lucy has A apples. How many apples does Julie have? "Fewer" version suggests wrong operation. Lucy has B fewer apples than Julie. Lucy has A apples. How many apples does Julie have? $A + B = \square$	"Fewer" version suggests operation. Lucy has B fewer apples than Julie. Julie has C apples. How many apples does Lucy have? "More" version suggests wrong operation. Julie has B more apples than Lucy. Julie has C apples. How many apples does Lucy have? $C - B = \square$ $\square + B = C$

Darker shading indicates the four Kindergarten problem subtypes. Grade 1 and 2 students work with all subtypes and variants. Unshaded (white) problems are the four difficult subtypes or variants that students should work with in Grade 1 but need not master until Grade 2. Adapted from CCSS, p. 88, which is based on *Mathematics Learning in Early Childhood: Paths Toward Excellence and Equity*, National Research Council, 2009, pp. 32–33.

[1] This can be used to show all decompositions of a given number, especially important for numbers within 10. Equations with totals on the left help children understand that = does not always mean "makes" or "results in" but always means "is the same number as." Such problems are not a problem subtype with one unknown, as is the Addend Unknown subtype to the right. These problems are a productive variation with two unknowns that give experience with finding all of the decompositions of a number and reflecting on the patterns involved.

[2] Either addend can be unknown; both variations should be included.

Topic A: Comparison Word Problems

Topic A

A Teaching Sequence Toward Mastery of Comparison Word Problems

Objective 1: Solve *compare with difference unknown* problem types.
(Lesson 1)

Objective 2: Solve *compare with bigger or smaller unknown* problem types.
(Lesson 2)

Lesson 1

Objective: Solve *compare with difference unknown* problem types.

Suggested Lesson Structure

- **Fluency Practice** (12 minutes)
- **Concept Development** (38 minutes)
- **Student Debrief** (10 minutes)
- **Total Time** **(60 minutes)**

Fluency Practice (12 minutes)

- Core Fluency Differentiated Practice Sets **1.OA.6** (5 minutes)
- Number Bond Addition and Subtraction **1.OA.6** (5 minutes)
- Happy Counting **1.NBT.1** (2 minutes)

Core Fluency Differentiated Practice Sets (5 minutes)

Materials: (S) Core Fluency Practice Sets

Note: Give the appropriate Practice Set to each student. Students who completed all questions correctly on their most recent Practice Set should be given the next level of difficulty. All other students should try to improve their scores on their current levels. Core Fluency Differentiated Practice Sets are used throughout this module.

Students complete as many problems as they can in 90 seconds. Assign a counting pattern and start number for early finishers, or have them practice make ten addition or subtraction on the back of their papers. Collect and correct any Practice Sets completed within the allotted time.

Number Bond Addition and Subtraction (5 minutes)

Materials: (S) Personal white board, die per pair

Note: Practice with missing addends and subtraction helps prepare students to solve comparison problems in today's Concept Development.

- Assign partners of equal ability.
- Allow partners to choose a number for their whole (within 10) and roll the die to determine one of the parts.

| A STORY OF UNITS | Lesson 1 1•6 |

- Both students write two addition and two subtraction sentences with a box representing the unknown number in each equation and solve for the missing number.
- Students exchange boards and check each other's work.

Happy Counting (2 minutes)

Note: In this module, students add and subtract within 100 and extend their counting and number writing skills to 120. Give students practice counting by ones and tens within 100. When Happy Counting by ones, spend more time changing directions where changes in tens occur, which is typically more challenging.

Happy Count by ones the regular way and the Say Ten way between 60 and 100. Then, Happy Count by tens, starting at a number with some ones (e.g., 78).

T:

T/S: 97   96   (pause)   97   98   (pause)   99   100   99   100   (etc.)

Concept Development (38 minutes)

Materials: (T) 4 ten-sticks, 2 charts with today's story problems (S) Personal math toolkit with 4 ten-sticks, personal white board

Note: Prepare two charts, one with the first story problem about Rose and another with the second story problem about Rose and Nikil. Save the second chart, with the solution, for tomorrow's lesson. Today's lesson objective is addressing word problems. Therefore, there is no separate Application Problem.

Gather students in the meeting area with their materials.

NOTES ON MULTIPLE MEANS OF REPRESENTATION:

Some students may find it helpful to use linking cubes to represent the problems. Students can use different color linking cubes for each part being represented and then draw the tape diagrams to match their concrete representations.

Problem 1: Model a *change unknown* problem with numerals within the tape rather than dots.

T: (Post the chart with the story problem.) Let's read this story problem together.

T/S: Rose wrote 8 letters to her friends. Her goal is to write 12 letters. How many more letters does she need to write to meet her goal?

T: Use a tape diagram to solve how many more letters Rose needs to write. You may also use your linking cubes to help draw and solve.

S: (Solve as the teacher circulates and notices various strategies.)

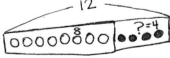

14   Lesson 1:   Solve *compare with difference unknown* problem types.

T: (Choose a student who used a tape diagram to solve. As the student shares, draw the tape diagram on the chart paper.)

S: I drew a rectangle around 8 circles to show how many letters Rose already wrote. Then, I drew a rectangle with a question mark because we need to find out how many more letters she needs to write. Then, I put arms from the first part to the end of the second part because I knew that she wants to write 12 letters. 8 + 4 = 12, so the answer is 4 letters.

T: Great. (Show a 12-stick of linking cubes made of 8 red and 4 yellow cubes.) I made a model of this story using linking cubes. Watch me as I draw my tape diagram only using numbers. Read the first sentence of the story problem.

S: Rose wrote 8 letters to her friends.

T: (Draw a tape, and label it *R*.) This represents the letters Rose wrote. What number should I write inside? (Point to the linking cubes.)

S: 8.

T: (Write 8 inside the tape.) Read the next sentence.

S: Her goal is to write 12 letters.

T: Is that a part of how many letters she wants to write, or is it the total letters she wants to write?

S: The total.

T: So, that means there are some more letters Rose needs to write. We just don't know how many more yet. (Draw another part, write in a question mark, and label it *M* as shown to the right. Point to the additional part of the linking cubes.)

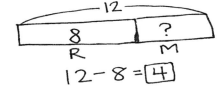

T: These two parts (point to each) make up the total of how many letters?

S: 12 letters.

T: (Draw the arms with 12, and then hold the linking cube stick at both ends, mimicking the arms drawn in the diagram.) What addition sentence helps find the missing part?

S: 8 + ___ = 12.

T: What is the subtraction number sentence to find the missing part?

S: 12 − 8 = 4.

T: How many more letters does Rose need to write?

S: 4 letters.

NOTES ON MULTIPLE MEANS OF REPRESENTATION:

To connect students' use of linking cubes to model the problem with the tape diagram, write the numbers for each part on stickers, and adhere the stickers to each part while drawing the tape diagram. A sticker with a question mark can be used to represent the unknown number.

Problem 2: Model a *compare with difference unknown* problem.

T: (Post the second chart with the next story problem.) Let's read another story problem together.

T/S: Rose wrote 8 letters. Nikil wrote 12 letters. How many more letters did Nikil write than Rose?

T: Partner A, using one color, make a stick of how many letters Rose wrote. Partner B, using a different color, make a stick to show the number of letters Nikil wrote. (Allow students time to make their sticks.)

T: Lay the two sticks down on the personal white board so we can compare them easily.

T: I see that many of you put your sticks side by side so that they are easier to compare. Let's all turn our sticks the same way so we can talk about them together. (Demonstrate by laying down the sticks horizontally on a personal white board, as shown on the right.) (Point to the 8-stick.) This stick represents whose letters?

S: Rose's.

T: (Label R on the personal white board as shown.) (Point to the 12-stick.) This stick represents...?

S: Nikil's letters.

T: (Label with N as shown.) Watch me as I use these cubes to help me draw my tape diagram to compare the number of letters Rose and Nikil wrote. (Write R.) How many letters did Rose write?

S: 8 letters.

T: (Draw a rectangle, and write 8 inside.)

T: (Write N in the next line.) How many letters did Nikil write?

S: 12 letters.

T: Will his tape, his part, be longer or shorter than Rose's tape, her part?

S: Longer!

T: Tell me when to stop when you think the length of the tape represents 12. (Begin drawing the tape.)

S: Stop!

T: (Stop at an appropriate length to represent 12, and complete the rectangle.) What number goes with this tape?

S: 12.

T: The question says, "How many more letters did Nikil write than Rose?" This tape (point to Rose's tape) represents 8, so this much of Nikil's tape is also 8. (Partition Nikil's tape with a dotted line, and write 8.) This part of Nikil's tape represents how many more letters he wrote. (Circle that part of Nikil's tape, and write a question mark as shown to the right.)

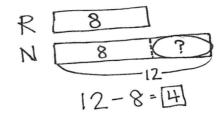

T: What is the total number of letters Nikil wrote?

S: 12 letters.

T: What is the part of Nikil's letters that are the same number as Rose's letters?

S: The 8 letters.

T: (Point to the question mark.) How many more letters did Nikil write than Rose? What can we do to figure out the unknown part? Turn and talk to your partner.

S: I compared the linking cubes we made and counted the extra cubes. I counted on. → There were 8, and I counted on 4 more to get to 12. There were 4 more cubes. → I thought 8 + ____ = 12. It's 4. → I used subtraction. I took away 8 from 12 and got 4.

T: If we count on 4 more from 8, we are adding 8 + 4 to get 12. If we cover up the 8 to see how many more letters he wrote, that is the same as taking away 8 from…?

S: 12.

T: What is 12 – 8?

S: 4.

T: How many more letters did Nikil write?

S: 4 letters.

T: I want you to see that we can use subtraction to compare the number of letters Rose and Nikil wrote.

T: Who wrote fewer letters?

S: Rose.

T: How do you know?

S: The tape diagram is shorter than Nikil's. → We know that Nikil wrote more, so Rose wrote fewer.

T: How many fewer letters did Rose write than Nikil? How do you know?

S: Four fewer letters! → Look at Rose's tape diagram. She needs 4 more to match Nikil's tape diagram. → Eight is 4 less than 12. → Nikil wrote 4 more letters, so Rose wrote 4 fewer letters. → Take away 8 from 12, and that tells you how many fewer letters Rose wrote.

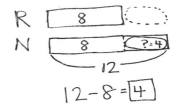

T: (Draw an invisible circle around the space after Rose's tape that would be where the additional letters would need to be for Rose to have the same number of letters as Nikil.) This part is the same length as Nikil's extra 4 letters. (In the image to the right, a dotted line is included to show where to demonstrate the invisible circle.)

Repeat the process with the following story problems. For each problem, ask students to use the linking cubes with their partners to represent the story. Guide them through drawing the double tape diagrams.

Tamra collected 9 seashells on the beach. Julio collected 11 seashells.

a. How many more seashells did Julio collect?
b. How many fewer seashells did Tamra collect?
c. How many seashells did Tamra and Julio collect? (This component provides a good contrast between the *comparison* problem type and a *put together* problem type.)

Lesson 1: Solve *compare with difference unknown* problem types.

A STORY OF UNITS

Lesson 1 1•6

Willie saw 13 leaping lizards at the park.
Fran saw 8 leaping lizards.

   a. How many more lizards did Willie see?
   b. How many fewer lizards did Fran see?
   c. How many lizards did Willie and Fran see?

Problem Set (10 minutes)

Students should do their personal best to complete the Problem Set within the allotted 10 minutes. Some problems do not specify a method for solving. This is an intentional reduction of scaffolding that invokes MP.5, Use Appropriate Tools Strategically. Students should solve these problems using the RDW approach used for Application Problems.

For some classes, it may be appropriate to modify the assignment by specifying which problems students should work on first. With this option, let the purposeful sequencing of the Problem Set guide the selections so that problems continue to be scaffolded. Balance word problems with other problem types to ensure a range of practice. Consider assigning incomplete problems for homework or at another time during the day.

Student Debrief (10 minutes)

Lesson Objective: Solve *compare with difference unknown* problem types.

The Student Debrief is intended to invite reflection and active processing of the total lesson experience.

Invite students to review their solutions for the Problem Set. They should check work by comparing answers with a partner before going over answers as a class. Look for misconceptions or misunderstandings that can be addressed in the Debrief. Guide students in a conversation to debrief the Problem Set and process the lesson.

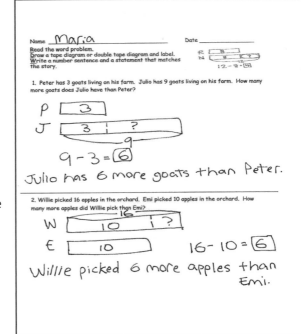

18   Lesson 1:   Solve *compare with difference unknown* problem types.

Any combination of the questions below may be used to lead the discussion.

- Look at Problem 1. Using the same story, how many fewer goats does Peter have than Julio? What do you notice about the answer to the question in the problem and this new question? Explain your thinking. How was setting up Problem 3 similar to and different from setting up Problems 1 and 2? What did you need to be sure to do? Why?
- When we know the total and just one of the parts, what strategy did we use to solve for the missing part?
- When two tapes are arranged one above the other like the ones we used today, we call that a *double tape diagram*. How does setting up our two tapes this way help you compare more easily?

Exit Ticket (3 minutes)

After the Student Debrief, instruct students to complete the Exit Ticket. A review of their work will help with assessing students' understanding of the concepts that were presented in today's lesson and planning more effectively for future lessons. The questions may be read aloud to the students.

Homework

Homework at the K–1 level is not a convention in all schools. In this curriculum, homework is an opportunity for additional practice of the content from the day's lesson. The teacher is encouraged, with the support of parents, administrators, and colleagues, to discern the appropriate use of homework for his or her students. Fluency exercises can also be considered as an alternative homework assignment.

Lesson 1: Solve *compare with difference unknown* problem types.

A STORY OF UNITS

Lesson 1 Core Fluency Practice Set A 1•6

Name _____   Date _____

My Addition Practice

1. 6 + 0 = ___	11. 7 + 1 = ___	21. 5 + 3 = ___
2. 0 + 6 = ___	12. ___ = 1 + 7	22. ___ = 5 + 4
3. 5 + 1 = ___	13. 3 + 3 = ___	23. 6 + 4 = ___
4. 1 + 5 = ___	14. 3 + 4 = ___	24. 4 + 6 = ___
5. 6 + 1 = ___	15. ___ = 3 + 5	25. ___ = 4 + 4
6. 1 + 6 = ___	16. 6 + 3 = ___	26. 3 + 4 = ___
7. 6 + 2 = ___	17. 7 + 3 = ___	27. 5 + 5 = ___
8. 5 + 2 = ___	18. ___ = 7 + 2	28. ___ = 4 + 5
9. 2 + 5 = ___	19. 2 + 7 = ___	29. 3 + 7 = ___
10. 2 + 4 = ___	20. 2 + 8 = ___	30. ___ = 3 + 6

Today I finished _____ problems.

I solved _____ problems correctly.

Lesson 1:   Solve *compare with difference unknown* problem types.

EUREKA MATH

A STORY OF UNITS　　　　　　Lesson 1 Core Fluency Practice Set B　　1•6

Name _____　　Date _____

My Missing Addend Practice

1. $6 + \_\_\_ = 6$
2. $0 + \_\_\_ = 6$
3. $5 + \_\_\_ = 6$
4. $4 + \_\_\_ = 6$
5. $0 + \_\_\_ = 7$
6. $6 + \_\_\_ = 7$
7. $1 + \_\_\_ = 7$
8. $7 + \_\_\_ = 8$
9. $1 + \_\_\_ = 8$
10. $6 + \_\_\_ = 8$

11. $3 + \_\_\_ = 6$
12. $4 + \_\_\_ = 8$
13. $10 = 5 + \_\_\_$
14. $5 + \_\_\_ = 9$
15. $5 + \_\_\_ = 7$
16. $8 = 5 + \_\_\_$
17. $5 + \_\_\_ = 9$
18. $8 + \_\_\_ = 10$
19. $7 + \_\_\_ = 10$
20. $10 = 6 + \_\_\_$

21. $4 + \_\_\_ = 7$
22. $7 = 3 + \_\_\_$
23. $2 + \_\_\_ = 7$
24. $2 + \_\_\_ = 8$
25. $9 = 2 + \_\_\_$
26. $2 + \_\_\_ = 10$
27. $10 = 3 + \_\_\_$
28. $3 + \_\_\_ = 9$
29. $4 + \_\_\_ = 9$
30. $10 = 4 + \_\_\_$

Today I finished _____ problems.

I solved _____ problems correctly.

Lesson 1:　　Solve *compare with difference unknown* problem types.

Name _____   Date _____

My Related Addition and Subtraction Practice

1. 5 + ___ = 6	11. 7 + ___ = 10	21. 4 + ___ = 8
2. 1 + ___ = 6	12. 10 − 7 = ___	22. 8 − 4 = ___
3. 6 − 1 = ___	13. 5 + ___ = 7	23. 4 + ___ = 7
4. 9 + ___ = 10	14. 7 − 5 = ___	24. 7 − 4 = ___
5. 1 + ___ = 10	15. 5 + ___ = 8	25. 5 + ___ = 9
6. 10 − 9 = ___	16. 8 − 5 = ___	26. 9 − 5 = ___
7. 5 + ___ = 10	17. 4 + ___ = 6	27. 6 + ___ = 9
8. 10 − 5 = ___	18. 6 − 4 = ___	28. 9 − 6 = ___
9. 8 + ___ = 10	19. 3 + ___ = 6	29. 4 + ___ = 7
10. 10 − 8 = ___	20. 6 − 3 = ___	30. 7 − 4 = ___

Today I finished _____ problems.

I solved _____ problems correctly.

My Subtraction Practice

Name _____ Date _____

1. 6 − 0 = ___
2. 6 − 1 = ___
3. 7 − 1 = ___
4. 8 − 1 = ___
5. 6 − 2 = ___
6. 7 − 2 = ___
7. 9 − 2 = ___
8. 10 − 10 = ___
9. 10 − 9 = ___
10. 10 − 7 = ___

11. 6 − 3 = ___
12. 7 − 3 = ___
13. 9 − 3 = ___
14. 10 − 8 = ___
15. 10 − 6 = ___
16. 10 − 4 = ___
17. 10 − 5 = ___
18. 7 − 6 = ___
19. 7 − 5 = ___
20. 6 − 4 = ___

21. 8 − 4 = ___
22. 8 − 3 = ___
23. 8 − 5 = ___
24. 9 − 5 = ___
25. 9 − 4 = ___
26. 7 − 3 = ___
27. 10 − 7 = ___
28. 9 − 7 = ___
29. 9 − 6 = ___
30. 8 − 6 = ___

Today I finished _____ problems.

I solved _____ problems correctly.

Name _____ Date _____

My Mixed Practice

1. 4 + 2 = ___	11. 2 + ___ = 6	21. 8 − 5 = ___
2. 2 + ___ = 6	12. 6 − 2 = ___	22. 3 + ___ = 8
3. 6 = 3 + ___	13. 6 − 4 = ___	23. 8 = ___ + 5
4. 2 + 5 = ___	14. 5 + ___ = 7	24. ___ + 2 = 9
5. 7 = 5 + ___	15. 7 − 5 = ___	25. 9 = ___ + 7
6. 4 + 3 = ___	16. 7 − 4 = ___	26. 9 − 2 = ___
7. 7 = ___ + 4	17. 7 − 3 = ___	27. 9 − 7 = ___
8. 8 = ___ + 4	18. 8 = 6 + ___	28. 9 − 6 = ___
9. 4 + 5 = ___	19. 8 − 2 = ___	29. 9 = ___ + 4
10. 9 = ___ + 4	20. 8 − 6 = ___	30. 9 − 6 = ___

Today I finished _____ problems.

I solved _____ problems correctly.

Name _____   Date _____

Read the word problem.
Draw a tape diagram or double tape diagram and label.
Write a number sentence and a statement that matches the story.

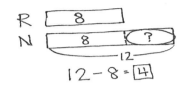

1. Peter has 3 goats living on his farm. Julio has 9 goats living on his farm. How many more goats does Julio have than Peter?

2. Willie picked 16 apples in the orchard. Emi picked 10 apples in the orchard. How many more apples did Willie pick than Emi?

Lesson 1:   Solve *compare with difference unknown* problem types.

3. Lee collected 13 eggs from the hens in the barn. Ben collected 18 eggs from the hens in the barn. How many fewer eggs did Lee collect than Ben?

4. Shanika did 14 cartwheels during recess. Kim did 20 cartwheels. How many more cartwheels did Kim do than Shanika?

Name _____ Date _____

Read the word problem.
Draw a tape diagram or double tape diagram and label.
Write a number sentence and a statement that matches the story.

R [8]
N [8][?]
      — 12 —
12 - 8 = [4]

Anton drove around the racetrack 12 times during the race. Rose drove around the racetrack 17 times. How many more times did Rose go around the racetrack than Anton?

Lesson 1: Solve *compare with difference unknown* problem types.

Name _____ Date _____

Read the word problem.
Draw a tape diagram or double tape diagram and label.
Write a number sentence and a statement that matches the story.

R [8]
N [8 | ?]
   — 12 —
12 − 8 = [4]

1. Fran donated 11 of her old books to the library. Darnel donated 8 of his old books to the library. How many more books did Fran donate than Darnel?

2. During recess, 7 students were reading books. There were 17 students playing on the playground. How many fewer students were reading books than playing on the playground?

3. Maria is 18 years old. Her brother Nikil is 12 years old. How much older is Maria than her brother Nikil?

4. It rained 15 days in the month of March. It rained 19 days in April. How many more days did it rain in April than in March?

Lesson 1: Solve *compare with difference unknown* problem types.

Lesson 2

Objective: Solve *compare with bigger or smaller unknown* problem types.

Suggested Lesson Structure

■ Fluency Practice   (12 minutes)
▢ Concept Development   (38 minutes)
■ Student Debrief   (10 minutes)
   Total Time   **(60 minutes)**

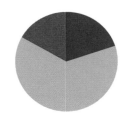

Fluency Practice (12 minutes)

- Core Fluency Differentiated Practice Sets **1.OA.6**   (5 minutes)
- Number Bond Addition and Subtraction **1.OA.6**   (5 minutes)
- Happy Counting **1.NBT.1**   (2 minutes)

Core Fluency Differentiated Practice Sets (5 minutes)

Materials: (S) Core Fluency Practice Sets (Lesson 1)

Note: Give the appropriate Practice Set to each student. Help students become aware of their improvement. After students finish today's Practice Sets, ask them to raise their hands if they tried a new level today or improved their score from the previous day.

Students complete as many problems as they can in 90 seconds. Assign a counting pattern and start number for early finishers, or have them practice make ten addition or subtraction on the back of their papers. Collect and correct any Practice Sets completed within the allotted time.

Number Bond Addition and Subtraction (5 minutes)

Materials: (S) Personal white board, die per pair

Note: Practice with missing addends and subtraction helps prepare students to solve comparison problems in today's Concept Development.

Conduct the activity as directed in Lesson 1.

Lesson 2

Happy Counting (2 minutes)

Note: In this module, students do addition and subtraction within 100 and extend their counting and number writing skills to 120. Give students practice counting by ones and tens within 100. When Happy Counting by ones, spend more time changing directions where changes in tens occur, which is typically more challenging.

Conduct the activity as directed in Lesson 1.

Concept Development (38 minutes)

Materials: (T) Chart with Lesson 1's tape diagram and Problem 2, chart with today's Problems 2 and 3, 4 ten-sticks (S) Personal math toolkit with 4 ten-sticks, personal white board

Note: Today's lesson objective is addressing word problems. Therefore, there is no separate Application Problem.

Gather students in the meeting area with their materials.

Problem 1

T: (Post the tape diagram from yesterday's Concept Development, Problem 2.)

T: What was the story that went with this tape diagram, in the last lesson?

S: Rose and Nikil both wrote letters. Rose wrote 8 letters, and Nikil wrote 12 letters. → How many more letters did Nikil write than Rose? → We also answered how many fewer letters Rose wrote than Nikil. → We also figured out how many letters Nikil and Rose wrote in all.

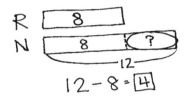

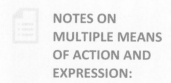

Nikil wrote 4 more letters than Rose.

T: Great! I have a new problem for you. (Point to the diagram as you speak.) Rose wrote 8 letters. Nikil wrote 4 more letters than Rose. How many letters did Nikil write? Turn and talk with your partner. (Wait as students discuss.)

T: If Rose wrote 8 letters, and Nikil wrote 4 more letters than Rose, how many letters did Nikil write?

S: 12 letters!

T: How do you know?

S: You have to add Rose's 8 letters and then 4 more. → You can look at the tape diagram on the chart. Nikil has the same 8 letters as Rose plus 4 more letters.

T: In the last lesson, you subtracted to find the difference between the two sets of letters. Is that what you did this time? Talk with a partner, and decide what number sentence you needed to use. (Wait as students discuss.)

NOTES ON MULTIPLE MEANS OF ACTION AND EXPRESSION:

If students struggle with word problems, consider using either smaller numbers or encouraging students to include circle representations for the objects, and then draw rectangles around the circles to create the tape diagrams.

S: We needed to add this time. → Eight letters plus 4 more letters is 12 letters. → 8 + 4 = 12.

Lesson 2: Solve *compare with bigger or smaller unknown* problem types.

Problem 2

T: Let's try another one. This time, use your linking cubes with a partner. Each of you will show linking cubes for your character.

T/S: Ben solved 6 math problems. Robin solved 4 more problems than Ben. How many problems did Robin solve?

T: Partner A, represent the problems Ben solved. Partner B, represent the problems Robin solved. Then, use your linking cubes to try to solve the problem together. (Circulate as students work to solve the problem. Remind them to read each sentence to recheck their work, making sure that their cubes match every part of the story.)

T: Let's draw a tape diagram to show what you just did. Who is this story about?

S: Ben and Robin.

T: (Write *B* and *R* to start a double tape diagram.) I like that most of you remembered to label your parts.

T: They each solved math problems. (Draw the same size rectangle next to each letter. This will help highlight the parts that are the same as well as the additional part that will be in Robin's tape.)

T: What do you notice about these two tapes?

S: They are the same size!

T: The same size tape means they solved the same amount of problems. Is this true?

S: No!

T: Who solved more problems?

S: Robin!

T: You are right! I'm going to add an extra part of tape next to Robin's to show that she solved more problems than Ben. (Draw.) How many more problems did Robin solve?

S: Four more problems.

T: Let's go back to our story. Read the first sentence.

S: Ben solved 6 math problems.

T: What information can I add to my double tape diagram?

S: Write 6 in Ben's tape!

T: Where else can I write in the 6? Turn and talk to your partner, and explain why.

S: Write 6 in the first part of Robin's tape. → It's the same size as Ben's tape, so it makes sense to put 6 there, too. → It makes sense to put 6 in Robin's first rectangle because the story says she solved 4 more than Ben. It has to show 4 more than 6 since 6 stands for how many problems Ben solved.

T: Great. (Write 6 in the first part of Robin's tape.) Does this match the linking cubes on your personal white board?

S: Yes!

T: If it doesn't, this is a good time to fix your model.

T: As I read each part of the story problem again, touch the part of the double tape model on your board that corresponds to what I'm saying.

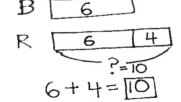

Lesson 2: Solve *compare with bigger or smaller unknown* problem types.

A STORY OF UNITS   Lesson 2 1•6

T/S: (Read each sentence, and have students point to the parts of their tape model.)

T: Write a number sentence that helped you find how many problems Robin solved.

S: 6 + 4 = 10.

T: How many problems did Robin solve?

S: Ten problems! (As students write 10 on the personal white board next to their model, add 10 to their double tape diagram as shown.)

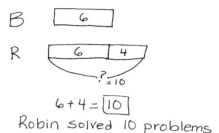

Problem 3

T: Let's read another story problem together.

T/S: Tamra found 12 ladybugs. Willie found 4 fewer ladybugs than Tamra. How many ladybugs did Willie find?

T: Who are the children in this story problem?

S: Tamra and Willie!

T: (Record *T* and *W* to begin a double tape diagram, and draw two equal size rectangles.)

T: Is it true that they found the same number of ladybugs?

S: No!

T: Who found *more* ladybugs? Read the story carefully again. Then, turn and talk to your partner.

S: Tamra. → It didn't say Tamra found more. But it said Willie found 4 *fewer* ladybugs. That means Tamra found *more*.

T: Great thinking! I need to add an extra tape, the "more tape," onto …?

S: Tamra's tape!

T: (Add an extra box.) How many more ladybugs did Tamra find than Willie?

S: 4 more ladybugs.

T: (Record 4 in the extra tape.) Let's read the first sentence of the story.

T/S: Tamra found 12 ladybugs.

T: Take a look at Tamra's tape. Turn and talk to your partner about where the 12 should go.

S: It should go inside the first part of the tape. → No. It should go outside like we did yesterday for Nikil's 12 letters. Twelve is the total number of ladybugs, so we need to put the arms around the entire tape for Tamra.

> **NOTES ON MULTIPLE MEANS OF REPRESENTATION:**
>
> Solving problems with the word *fewer* can be difficult, especially for English language learners. When solving problems of this type, teach students to always focus on "who has more." For example, after reading the problem, before solving, have students look at who has fewer and who has more. Establishing this before solving makes sure students really understand how to solve this problem type.

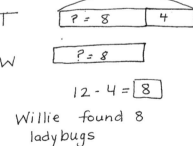

Lesson 2: Solve *compare with bigger or smaller unknown* problem types.

T:	Hmm, let's try the first idea and see. (Write 12 in the first tape.) According to Tamra's tape now, did she find 12 ladybugs?
S:	No. It looks like she found 16 ladybugs.
T:	You are right. Is 12 the total amount of ladybugs Tamra found or just a part?
S:	The total.
T:	Let's try the other suggestion.
T:	(Make a bracket with 12 for Tamra's tape.) Does this show that Tamra found a total of 12 ladybugs?
S:	Yes!
T:	Read the next sentence.
S:	Willie found 4 fewer ladybugs than Tamra.
T:	Did we show that in our double tape diagram?
S:	Yes!
T:	Read the last part of our story problem.
S:	How many ladybugs did Willie find?
T:	(Record a question mark in Willie's tape.) Look at Willie's tape. What do you notice about the size of the tape?
S:	It's the same as the first part of Tamra's tape.
T:	If we find out what the missing part for Tamra's tape is, then we are also finding out…?
S:	Willie's tape.
T:	How can we find this missing part of Tamra's tape? Turn and talk to your partner.
S:	I did 4 + ___ = 12. The answer is 8. → I used subtraction to find the missing part. 12 − 4 = 8. The missing part is 8.
T:	Great. If this part is 8 (fill in the 8 to complete Tamra's tape), then what else is 8?
S:	Willie's tape!
T:	So, how many ladybugs did Willie find?
S:	8 ladybugs!

Repeat the process by using the following story problems. For each problem, guide students through drawing the double tape diagram.

- Shanika used 11 blocks to build a house. Julio used 5 more blocks than Shanika. How many blocks did Julio use?
- Darnel caught 10 fewer fish than Fran. Fran caught 16 fish. How many fish did Darnel catch?
- Maria found 9 flowers in the garden. Kiana found 12 flowers. How many more flowers did Kiana find than Maria?

Problem Set (10 minutes)

Students should do their personal best to complete the Problem Set within the allotted 10 minutes. For some classes, it may be appropriate to modify the assignment by specifying which problems they work on first. Some problems do not specify a method for solving. Students should solve these problems using the RDW approach used for Application Problems.

A STORY OF UNITS

Lesson 2   1•6

Student Debrief (10 minutes)

Lesson Objective: Solve *compare with bigger or smaller unknown* problem types.

The Student Debrief is intended to invite reflection and active processing of the total lesson experience.

Invite students to review their solutions for the Problem Set. They should check work by comparing answers with a partner before going over answers as a class. Look for misconceptions or misunderstandings that can be addressed in the Debrief. Guide students in a conversation to debrief the Problem Set and process the lesson.

Any combination of the questions below may be used to lead the discussion.

- Look at Problems 1 and 2. How was drawing Nikil's tape and Emi's tape different? Explain why this is so.
- How was setting up the tape diagram from Problem 3 different from Problem 1?
- Explain to your partner how you solved Problem 6.
- In which problem were you able to use your doubles or doubles plus 1 facts to solve?
- How did working on number bond addition and subtraction in today's fluency activity help you with solving today's story problems?

Exit Ticket (3 minutes)

After the Student Debrief, instruct students to complete the Exit Ticket. A review of their work will help with assessing students' understanding of the concepts that were presented in today's lesson and planning more effectively for future lessons. The questions may be read aloud to the students.

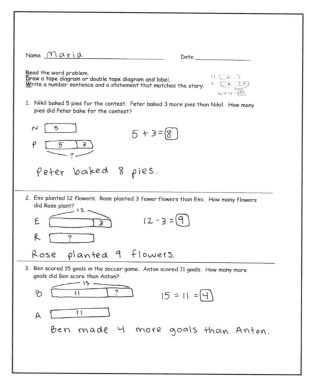

Lesson 2:   Solve *compare with bigger or smaller unknown* problem types.

35

Name _____   Date _____

Read the word problem.
Draw a tape diagram or double tape diagram and label.
Write a number sentence and a statement that matches the story.

N [6]
R [6 | 4]
   ?=10
6 + 4 = [10]

1. Nikil baked 5 pies for the contest. Peter baked 3 more pies than Nikil. How many pies did Peter bake for the contest?

2. Emi planted 12 flowers. Rose planted 3 fewer flowers than Emi. How many flowers did Rose plant?

3. Ben scored 15 goals in the soccer game. Anton scored 11 goals. How many more goals did Ben score than Anton?

4. Kim grew 12 roses in a garden. Fran grew 6 fewer roses than Kim. How many roses did Fran grow in the garden?

5. Maria has 4 more fish in her tank than Shanika. Shanika has 16 fish. How many fish does Maria have in her tank?

6. Lee has 11 board games. Lee has 5 more board games than Darnel. How many board games does Darnel have?

A STORY OF UNITS   Lesson 2 Exit Ticket 1•6

Name _____   Date _____

Read the word problem.
Draw a tape diagram or double tape diagram and label.
Write a number sentence and a statement that matches the story.

N [6]
R [6 | 4]
    ?=10
6 + 4 = [10]

Tamra decorated 13 cookies. Kiana decorated 5 fewer cookies than Tamra. How many cookies did Kiana decorate?

38   Lesson 2:   Solve *compare with bigger or smaller unknown* problem types.

A STORY OF UNITS **Lesson 2 Homework 1•6**

Name _____ Date _____

Read the word problem.
Draw a tape diagram or double tape diagram and label.
Write a number sentence and a statement that matches the story.

N [6]
R [6 | 4]
  ?=10
6 + 4 = [10]

1. Kim went to 15 baseball games this summer. Julio went to 10 baseball games. How many more games did Kim go to than Julio?

2. Kiana picked 14 strawberries at the farm. Tamra picked 5 fewer strawberries than Kiana. How many strawberries did Tamra pick?

3. Willie saw 7 reptiles at the zoo. Emi saw 4 more reptiles at the zoo than Willie. How many reptiles did Emi see at the zoo?

Lesson 2: Solve *compare with bigger or smaller unknown* problem types.

4. Peter jumped into the swimming pool 6 times more than Darnel. Darnel jumped in 9 times. How many times did Peter jump into the swimming pool?

5. Rose found 16 seashells on the beach. Lee found 6 fewer seashells than Rose. How many seashells did Lee find on the beach?

6. Shanika got 12 cards in the mail. Nikil got 5 more cards than Shanika. How many cards did Nikil get?

A STORY OF UNITS

Mathematics Curriculum

GRADE 1 • MODULE 6

Topic B
Numbers to 120

1.NBT.1, 1.NBT.2a, 1.NBT.2c, 1.NBT.3, 1.NBT.5

Focus Standards:	1.NBT.1	Count to 120, starting at any number less than 120. In this range, read and write numerals and represent a number of objects with a written numeral.
	1.NBT.2	Understand that the two digits of a two-digit number represent amounts of tens and ones. Understand the following as special cases:
		a. 10 can be thought of as a bundle of ten ones—called a "ten."
		c. The numbers 10, 20, 30, 40, 50, 60, 70, 80, 90 refer to one, two, three, four, five, six, seven, eight, or nine tens (and 0 ones).
	1.NBT.3	Compare two two-digit numbers based on meanings of the tens and ones digits, recording the results of comparisons with the symbols >, =, and <.
	1.NBT.5	Given a two-digit number, mentally find 10 more or 10 less than the number, without having to count; explain the reasoning used.
Instructional Days:	7	
Coherence -Links from:	G1–M4	Place Value, Comparison, Addition and Subtraction of Numbers to 40
-Links to:	G2–M3	Place Value, Counting, and Comparison of Numbers to 1,000

Topic B extends students' use of counting sequences and understanding of tens and ones to numbers up to and including 120.

In Lesson 3, students apply their understanding of tens and ones to two-digit numbers greater than 40. Students count by tens and then extra ones to efficiently count large groups of objects. They then use the place value chart to record quantities as tens and ones as well as by their traditional number (**1.NBT.2**).

In Lesson 4, students connect this understanding with its application to addition sentences. Students recognize that numbers such as 67 can be interpreted as 6 tens 7 ones and that the units can be combined to find the total: 60 + 7 = 67. This work of decomposing and composing 67 into its tens and ones supports the work students do in Topic C, as they decompose two-digit numbers before adding to another two-digit number.

Students continue to consider tens and ones in Lesson 5 when they identify 10 more, 10 less, 1 more, and 1 less than any two-digit number (**1.NT.5**). This work helps students attend to the parts within a two-digit number, a skill that is critical to adding two-digit numbers within 100. Students recognize that when looking at a number such as 37, they focus on the tens place when adding or subtracting 10 and on the ones place when adding or subtracting 1. Students also explore numbers such as 89, where adding 1 more creates another ten.

Topic B: Numbers to 120

41

During Lesson 6, students practice comparing numbers using the symbols >, =, and < (**1.NBT.3**). They compare numbers such as 65 and 75, as well as numbers in various unit form combinations such as 7 tens 5 ones, 5 ones 7 tens, and 6 tens 15 ones. Through these explorations, students consider ways that each number can be decomposed and recomposed.

In Lesson 7, students work with the counting sequence to 120 (**1.NBT.1**). After counting from 78 to 120, students use Hide Zero cards to build numbers from 100 to 120. Their strong familiarity with counting from 0 to 20 and back is then related to the sequence from 100 to 120, helping students recognize that their prior knowledge can help them succeed at this new level.

Lesson 8 continues the use of the Hide Zero cards, as students use 5-group cards of 10 to write numbers within place value charts. Students represent 100 as 10 tens and then represent 101 as 10 tens and 1 one. This work with the unit form of numbers to 120 supports students' understanding of the written numerals 101 through 109, which are the most challenging to write (**1.NBT.1**).

Following students' work with the unit form of numbers to 120, students then represent a number of objects in Lesson 9, presented concretely and pictorially, with the written numeral (**1.NBT.1**).

A Teaching Sequence Toward Mastery of Numbers to 120
Objective 1: Use the place value chart to record and name tens and ones within a two-digit number up to 100.   (Lesson 3)
Objective 2: Write and interpret two-digit numbers to 100 as addition sentences that combine tens and ones.   (Lesson 4)
Objective 3: Identify 10 more, 10 less, 1 more, and 1 less than a two-digit number within 100.   (Lesson 5)
Objective 4: Use the symbols >, =, and < to compare quantities and numerals to 100.   (Lesson 6)
Objective 5: Count and write numbers to 120. Use Hide Zero cards to relate numbers 0 to 20 to 100 to 120.   (Lesson 7)
Objective 6: Count to 120 in unit form using only tens and ones. Represent numbers to 120 as tens and ones on the place value chart.   (Lesson 8)
Objective 7: Represent up to 120 objects with a written numeral.   (Lesson 9)

Lesson 3

Objective: Use the place value chart to record and name tens and ones within a two-digit number up to 100.

Suggested Lesson Structure

- Application Problem (5 minutes)
- Fluency Practice (15 minutes)
- Concept Development (30 minutes)
- Student Debrief (10 minutes)

Total Time **(60 minutes)**

Application Problem (5 minutes)

Tamra has 4 more goldfish than Peter. Peter has 10 goldfish. How many goldfish does Tamra have?

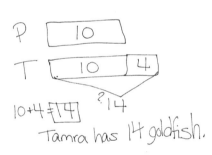

Note: Throughout Module 6, the Application Problem comes before the Fluency Practice so that the core fluency can move directly into the operations with two-digit numbers. Today's Application Problem continues students' practice with the *compare with bigger unknown* problem type, which was part of Lesson 2's objective.

Fluency Practice (15 minutes)

- Grade 1 Core Fluency Sprint **1.OA.6**   (10 minutes)
- Subtraction with Cards **1.OA.6**   (5 minutes)

Grade 1 Core Fluency Sprint (10 minutes)

Materials: (S) Core Fluency Sprints

Note: Choose an appropriate Sprint based on the needs of the class. For today's movement-counting between Sprints A and B, consider practicing counting the Say Ten way to prepare students for today's lesson. Follow the suggested counting pattern: Count by ones from 37 to 52 and back, and then count by tens from 87 to 107 and back.

Core Fluency Sprint List:

- Core Addition Sprint 1 (targeting core addition and missing addends)
- Core Addition Sprint 2 (targeting the most challenging addition within 10 and beyond 10)
- Core Subtraction Sprint (targeting core subtraction)
- Core Fluency Sprint: Totals of 5, 6, and 7 (developing understanding of the relationship between addition and subtraction)
- Core Fluency Sprint: Totals of 8, 9, and 10 (developing understanding of the relationship between addition and subtraction)

NOTES ON MULTIPLE MEANS OF REPRESENTATION:

Differentiating Sprints for students helps meet the needs of the class. Adjust them to suit specific learning needs so students feel successful and do not show frustration while completing them.

Subtraction with Cards (5 minutes)

Materials: (S) 1 pack of numeral cards 0–10 per set of partners (Fluency Template)

Note: This review activity strengthens students' ability to subtract within 10, which supports their work decomposing numbers in future lessons within the module.

- Students combine their digit cards and place them facedown between them.
- Each partner flips over two cards and subtracts the smaller number from the larger one.
- The partner with the smallest difference keeps the cards played by both players in that round.
- If the differences are equal, the cards are set aside, and the winner of the next round keeps the cards from both rounds.
- A player wins by having the most cards when the time is up.

Concept Development (30 minutes)

Materials: (T) Hide Zero cards (Template 1), chart paper (S) 4 ten-sticks from personal math toolkit, personal white board, place value chart (Template 2)

Students sit at their desks with their materials.

T: (Show 47 using Hide Zero cards.) What number am I showing?
S: 47.
T: When I pull apart these Hide Zero cards, 47 will be in two parts. What will they be?
S: 40 and 7.
T: (Write 40 and 7 on the board.) You're right! Explain to your partner why we don't see 40 but just the digit 4. (Listen as partners explain their thinking to each other.)
S: When you pull apart the cards, you'll see the 0 hiding behind 7. → 4 stands for 40 because it's in the tens place. 7 stands for just 7 ones.

A STORY OF UNITS   Lesson 3   1•6

T: (Pull apart 47 into 40 and 7.) You are right! Show me 47 using quick ten drawings. Count out each ten, and add on each of the ones the Say Ten way as you draw them.
S: 1 ten, 2 tens, 3 tens, 4 tens, 4 tens 1, 4 tens 2, …, 4 tens 7.
T: How many tens did you draw?
S: 4 tens.
T: How many ones did you draw?
S: 7 ones.
T: Let's fill in the place value chart. How many tens are in 47?
S: 4 tens.
T: Let's write 4 in the…?
S: Tens place. (Fill in 4.)
T: How many ones are in 47?
S: 7 ones.
T: Let's write 7 in the…?
S: Ones place. (Fill in 7.)

> **NOTES ON MULTIPLE MEANS OF ENGAGEMENT:**
>
> Provide challenging extensions for students. Give clues with tens and ones, and have students guess the number. For example, "What number is made up of …?"
>
> 2 tens and 23 ones, 6 tens and 35 ones, 1 ten and 47 ones, 9 tens and 14 ones, etc.

Repeat the process with the following suggested sequence: 57, 67, 86, 68, 95, and 100.

MP.4
T: (Write 64 on the place value chart.) What does the digit 6 stand for?
S: 6 tens.
T: 6 tens is the same as…?
S: 60.
T: What does the digit 4 stand for?
S: 4 ones.
T: What is 6 tens and 4 ones, or 60 and 4?
S: 64.

Repeat the process using the following sequence: 74, 84, 93, 73, 65, 56, 79, 97, and 100.

Problem Set (10 minutes)

Students should do their personal best to complete the Problem Set within the allotted 10 minutes. For some classes, it may be appropriate to modify the assignment by specifying which problems they work on first. Some problems do not specify a method for solving. Students should solve these problems using the RDW approach used for Application Problems.

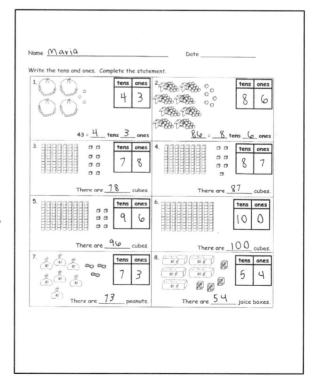

Lesson 3: Use the place value chart to record and name tens and ones within a two-digit number up to 100.

45

A STORY OF UNITS

Lesson 3 1•6

Student Debrief (10 minutes)

Lesson Objective: Use the place value chart to record and name tens and ones within a two-digit number up to 100.

The Student Debrief is intended to invite reflection and active processing of the total lesson experience.

Invite students to review their solutions for the Problem Set. They should check work by comparing answers with a partner before going over answers as a class. Look for misconceptions or misunderstandings that can be addressed in the Debrief. Guide students in a conversation to debrief the Problem Set and process the lesson.

Any combination of the questions below may be used to lead the discussion.

- Look at your answers for Problems 1 and 7. What is the difference between these two numbers? Explain how you know.
- For Problem 3, a student said there are 87 cubes. Is he correct? How can you help this student so he understands place value correctly?
- Using a quick ten drawing or your Hide Zero cards, explain how you solved Problem 9(j).
- Look at Problem 9(b). What must we add to 46 to get 5 tens and 0 ones?
- Think about the movement-counting we did between our two Sprints today. How can counting the Say Ten way help you think about the tens and ones in two-digit numbers? Use an example as you share your explanation.
- Look at your Application Problem. How did you solve the problem? Which problem from yesterday is this problem most like?

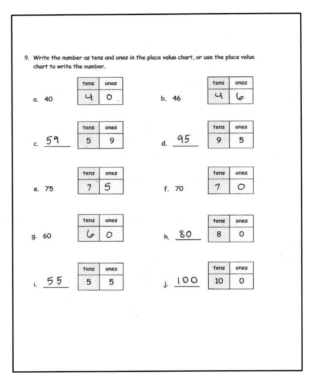

Exit Ticket (3 minutes)

After the Student Debrief, instruct students to complete the Exit Ticket. A review of their work will help with assessing students' understanding of the concepts that were presented in today's lesson and planning more effectively for future lessons. The questions may be read aloud to the students.

Lesson 3: Use the place value chart to record and name tens and ones within a two-digit number up to 100.

A

Lesson 3 Core Addition Sprint 1

Name _____ Date _____

Number Correct: _____

*Write the unknown number. Pay attention to the symbols.

1.	4 + 1 = ____	16.	4 + 3 = ____
2.	4 + 2 = ____	17.	____ + 4 = 7
3.	4 + 3 = ____	18.	7 = ____ + 4
4.	6 + 1 = ____	19.	5 + 4 = ____
5.	6 + 2 = ____	20.	____ + 5 = 9
6.	6 + 3 = ____	21.	9 = ____ + 4
7.	1 + 5 = ____	22.	2 + 7 = ____
8.	2 + 5 = ____	23.	____ + 2 = 9
9.	3 + 5 = ____	24.	9 = ____ + 7
10.	5 + ____ = 8	25.	3 + 6 = ____
11.	8 = 3 + ____	26.	____ + 3 = 9
12.	7 + 2 = ____	27.	9 = ____ + 6
13.	7 + 3 = ____	28.	4 + 4 = ____ + 2
14.	7 + ____ = 10	29.	5 + 4 = ____ + 3
15.	____ + 7 = 10	30.	____ + 7 = 3 + 6

Lesson 3: Use the place value chart to record and name tens and ones within a two-digit number up to 100.

A STORY OF UNITS Lesson 3 Core Addition Sprint 1 1•6

B

Name _____ Date _____

Number Correct:

*Write the unknown number. Pay attention to the symbols.

1.	5 + 1 = ____	16.	2 + 4 = ____
2.	5 + 2 = ____	17.	____ + 4 = 6
3.	5 + 3 = ____	18.	6 = ____ + 4
4.	4 + 1 = ____	19.	3 + 4 = ____
5.	4 + 2 = ____	20.	____ + 3 = 7
6.	4 + 3 = ____	21.	7 = ____ + 4
7.	1 + 3 = ____	22.	4 + 5 = ____
8.	2 + 3 = ____	23.	____ + 4 = 9
9.	3 + 3 = ____	24.	9 = ____ + 5
10.	3 + ____ = 6	25.	2 + 6 = ____
11.	____ + 3 = 6	26.	____ + 6 = 9
12.	5 + 2 = ____	27.	9 = ____ + 2
13.	5 + 3 = ____	28.	3 + 3 = ____ + 4
14.	5 + ____ = 8	29.	3 + 4 = ____ + 5
15.	____ + 3 = 8	30.	____ + 6 = 2 + 7

A STORY OF UNITS

Lesson 3 Core Addition Sprint 2   1•6

A

Name _____ Date _____

Number Correct: _____

*Write the unknown number. Pay attention to the equal sign.

1.	5 + 2 = ____	16.	____ = 5 + 4
2.	6 + 2 = ____	17.	____ = 4 + 5
3.	7 + 2 = ____	18.	6 + 3 = ____
4.	4 + 3 = ____	19.	3 + 6 = ____
5.	5 + 3 = ____	20.	____ = 2 + 6
6.	6 + 3 = ____	21.	2 + 7 = ____
7.	____ = 6 + 2	22.	____ = 3 + 4
8.	____ = 2 + 6	23.	3 + 6 = ____
9.	____ = 7 + 2	24.	____ = 4 + 5
10.	____ = 2 + 7	25.	3 + 4 = ____
11.	____ = 4 + 3	26.	13 + 4 = ____
12.	____ = 3 + 4	27.	3 + 14 = ____
13.	____ = 5 + 3	28.	3 + 6 = ____
14.	____ = 3 + 5	29.	13 + ____ = 19
15.	____ = 3 + 4	30.	19 = ____ + 16

Lesson 3:   Use the place value chart to record and name tens and ones within a two-digit number up to 100.

A STORY OF UNITS Lesson 3 Core Addition Sprint 2 **1•6**

B

Name _____   Date _____

Number Correct: _____

*Write the unknown number. Pay attention to the equal sign.

1.	4 + 3 = ____	16.	____ = 6 + 3
2.	5 + 3 = ____	17.	____ = 3 + 6
3.	6 + 3 = ____	18.	5 + 4 = ____
4.	6 + 2 = ____	19.	4 + 5 = ____
5.	7 + 2 = ____	20.	____ = 2 + 7
6.	5 + 4 = ____	21.	2 + 6 = ____
7.	____ = 4 + 3	22.	____ = 3 + 4
8.	____ = 3 + 4	23.	4 + 5 = ____
9.	____ = 5 + 3	24.	____ = 3 + 6
10.	____ = 3 + 5	25.	2 + 7 = ____
11.	____ = 6 + 2	26.	12 + 7 = ____
12.	____ = 2 + 6	27.	2 + 17 = ____
13.	____ = 7 + 2	28.	4 + 5 = ____
14.	____ = 2 + 7	29.	14 + ____ = 19
15.	____ = 7 + 2	30.	19 = ____ + 15

Lesson 3: Use the place value chart to record and name tens and ones within a two-digit number up to 100.

EUREKA MATH

A

Name _____ **Date** _____

Number Correct: _____

*Write the unknown number. Pay attention to the symbols.

1.	6 − 1 = ____	16.	8 − 2 = ____
2.	6 − 2 = ____	17.	8 − 6 = ____
3.	6 − 3 = ____	18.	7 − 3 = ____
4.	10 − 1 = ____	19.	7 − 4 = ____
5.	10 − 2 = ____	20.	8 − 4 = ____
6.	10 − 3 = ____	21.	9 − 4 = ____
7.	7 − 2 = ____	22.	9 − 5 = ____
8.	8 − 2 = ____	23.	9 − 6 = ____
9.	9 − 2 = ____	24.	9 − ____ = 6
10.	7 − 3 = ____	25.	9 − ____ = 2
11.	8 − 3 = ____	26.	2 = 8 − ____
12.	10 − 3 = ____	27.	2 = 9 − ____
13.	10 − 4 = ____	28.	10 − 7 = 9 − ____
14.	9 − 4 = ____	29.	9 − 5 = ____ − 3
15.	8 − 4 = ____	30.	____ − 6 = 9 − 7

Lesson 3: Use the place value chart to record and name tens and ones within a two-digit number up to 100.

A STORY OF UNITS   **Lesson 3 Core Subtraction Sprint** 1•6

B

Number Correct: _____

Name _____   Date _____

*Write the unknown number. Pay attention to the symbols.

1.	5 − 1 = ____	16.	6 − 2 = ____
2.	5 − 2 = ____	17.	6 − 4 = ____
3.	5 − 3 = ____	18.	8 − 3 = ____
4.	10 − 1 = ____	19.	8 − 5 = ____
5.	10 − 2 = ____	20.	8 − 6 = ____
6.	10 − 3 = ____	21.	9 − 3 = ____
7.	6 − 2 = ____	22.	9 − 6 = ____
8.	7 − 2 = ____	23.	9 − 7 = ____
9.	8 − 2 = ____	24.	9 − ____ = 5
10.	6 − 3 = ____	25.	9 − ____ = 4
11.	7 − 3 = ____	26.	4 = 8 − ____
12.	8 − 3 = ____	27.	4 = 9 − ____
13.	5 − 4 = ____	28.	10 − 8 = 9 − ____
14.	6 − 4 = ____	29.	8 − 6 = ____ − 7
15.	7 − 4 = ____	30.	____ − 4 = 9 − 6

Lesson 3: Use the place value chart to record and name tens and ones within a two-digit number up to 100.

EUREKA MATH

A STORY OF UNITS   Core Fluency Sprint: Totals of 5, 6, & 7   **1•6**

A

Name _____ Date _____

Number Correct: ____

*Write the unknown number. Pay attention to the symbols.

1.	2 + 3 = ____	16.	3 + 3 = ____
2.	3 + ____ = 5	17.	6 − 3 = ____
3.	5 − 3 = ____	18.	6 = ____ + 3
4.	5 − 2 = ____	19.	2 + 5 = ____
5.	____ + 2 = 5	20.	5 + ____ = 7
6.	1 + 5 = ____	21.	7 − 2 = ____
7.	1 + ____ = 6	22.	7 − 5 = ____
8.	6 − 1 = ____	23.	7 = ____ + 5
9.	6 − 5 = ____	24.	3 + 4 = ____
10.	____ + 5 = 6	25.	4 + ____ = 7
11.	4 + 2 = ____	26.	7 − 4 = ____
12.	2 + ____ = 6	27.	7 = ____ + 3
13.	6 − 2 = ____	28.	3 = 7 − ____
14.	6 − 4 = ____	29.	7 − 5 = ____ − 4
15.	____ + 4 = 6	30.	____ − 3 = 7 − 4

EUREKA MATH™   Lesson 3:   Use the place value chart to record and name tens and ones within a two-digit number up to 100.

A STORY OF UNITS — Core Fluency Sprint: Totals of 5, 6, & 7   1•6

B

Name _____   Date _____

Number Correct:

*Write the unknown number. Pay attention to the symbols.

1.	1 + 4 = ____	16.	3 + 3 = ____
2.	4 + ____ = 5	17.	6 − 3 = ____
3.	5 − 4 = ____	18.	6 = ____ + 3
4.	5 − 1 = ____	19.	2 + 4 = ____
5.	____ + 1 = 5	20.	4 + ____ = 6
6.	7 + 2 = ____	21.	6 − 2 = ____
7.	5 + ____ = 7	22.	6 − 4 = ____
8.	7 − 2 = ____	23.	6 = ____ + 4
9.	7 − 5 = ____	24.	3 + 4 = ____
10.	____ + 2 = 7	25.	4 + ____ = 7
11.	1 + 5 = ____	26.	7 − 4 = ____
12.	1 + ____ = 6	27.	7 = ____ + 4
13.	6 − 1 = ____	28.	4 = 7 − ____
14.	6 − 5 = ____	29.	6 − 4 = ____ − 5
15.	____ + 5 = 6	30.	____ − 4 = 7 − 3

Lesson 3:   Use the place value chart to record and name tens and ones within a two-digit number up to 100.

A Story of Units

Core Fluency Sprint: Totals of 8, 9, & 10 1•6

A

Name _____

Number Correct: _____

Date _____

*Write the unknown number. Pay attention to the symbols.

1.	5 + 5 = ____	16.	2 + 6 = ____
2.	5 + ____ = 10	17.	8 = 6 + ____
3.	10 − 5 = ____	18.	8 − 2 = ____
4.	9 + 1 = ____	19.	2 + 7 = ____
5.	1 + ____ = 10	20.	9 = 7 + ____
6.	10 − 1 = ____	21.	9 − 7 = ____
7.	10 − 9 = ____	22.	8 = ____ + 2
8.	____ + 9 = 10	23.	8 − 6 = ____
9.	1 + 8 = ____	24.	3 + 6 = ____
10.	8 + ____ = 9	25.	9 = 6 + ____
11.	9 − 1 = ____	26.	9 − 6 = ____
12.	9 − 8 = ____	27.	9 = ____ + 3
13.	____ + 1 = 9	28.	3 = 9 − ____
14.	4 + 4 = ____	29.	9 − 5 = ____ − 6
15.	8 − 4 = ____	30.	____ − 7 = 8 − 6

EUREKA MATH

Lesson 3: Use the place value chart to record and name tens and ones within a two-digit number up to 100.

A STORY OF UNITS   Core Fluency Sprint: Totals of 8, 9, & 10   **1•6**

B

Name _____ Date _____

Number Correct: _____

*Write the unknown number. Pay attention to the symbols.

1.	9 + 1 = ____	16.	3 + 5 = ____
2.	1 + ____ = 10	17.	8 = 5 + ____
3.	10 − 1 = ____	18.	8 − 3 = ____
4.	10 − 9 = ____	19.	2 + 6 = ____
5.	____ + 9 = 10	20.	8 = 6 + ____
6.	1 + 7 = ____	21.	8 − 6 = ____
7.	7 + ____ = 8	22.	2 + 7 = ____
8.	8 − 1 = ____	23.	9 = ____ + 2
9.	8 − 7 = ____	24.	9 − 7 = ____
10.	____ + 1 = 8	25.	4 + 5 = ____
11.	2 + 8 = ____	26.	9 = 5 + ____
12.	2 + ____ = 10	27.	9 − 5 = ____
13.	10 − 2 = ____	28.	5 = 9 − ____
14.	10 − 8 = ____	29.	9 − 6 = ____ − 5
15.	____ + 8 = 10	30.	____ − 6 = 9 − 7

Lesson 3: Use the place value chart to record and name tens and ones within a two-digit number up to 100.

A STORY OF UNITS　　　Lesson 3 Problem Set 1•6

Name _____ Date _____

Write the tens and ones. Complete the statement.

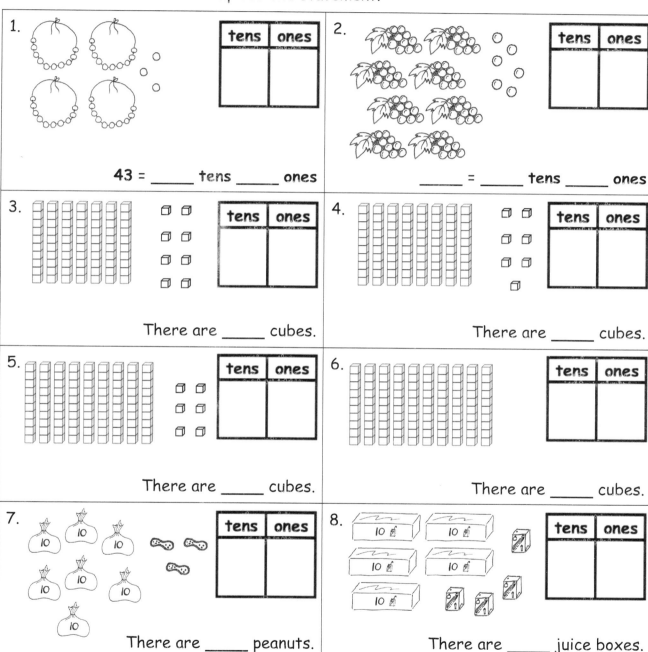

Lesson 3: Use the place value chart to record and name tens and ones within a two-digit number up to 100.

9. Write the number as tens and ones in the place value chart, or use the place value chart to write the number.

a. 40 | tens | ones |
|---|---|
| | |

b. 46 | tens | ones |
|---|---|
| | |

c. ___ | tens | ones |
|---|---|
| 5 | 9 |

d. ___ | tens | ones |
|---|---|
| 9 | 5 |

e. 75 | tens | ones |
|---|---|
| | |

f. 70 | tens | ones |
|---|---|
| | |

g. 60 | tens | ones |
|---|---|
| | |

h. ___ | tens | ones |
|---|---|
| 8 | 0 |

i. ___ | tens | ones |
|---|---|
| 5 | 5 |

j. ___ | tens | ones |
|---|---|
| 10 | 0 |

A STORY OF UNITS   Lesson 3 Exit Ticket 1•6

Name _____   Date _____

1. Write the tens and ones. Complete the statement.

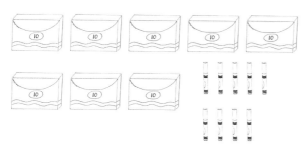

tens	ones

There are _____ markers.

2. Write the number as tens and ones in the place value chart, or use the place value chart to write the number.

a. 90

tens	ones

b. _____

tens	ones
8	7

Lesson 3: Use the place value chart to record and name tens and ones within a two-digit number up to 100.

A STORY OF UNITS

Lesson 3 Homework 1•6

Name _____ Date _____

Write the tens and ones. Complete the statement.

1.
tens	ones

52 = _____ tens _____ ones

2.
tens	ones

_____ = _____ tens _____ ones

3.
tens	ones

There are _____ cubes.

4.
tens	ones

There are _____ cubes.

5.
tens	ones

There are _____ cubes.

6.
tens	ones

There are _____ cubes.

7.
tens	ones

There are _____ carrots.

8.
tens	ones

There are _____ markers.

Lesson 3: Use the place value chart to record and name tens and ones within a two-digit number up to 100.

A STORY OF UNITS — Lesson 3 Homework 1•6

9. Write the number as tens and ones in the place value chart, or use the place value chart to write the number.

a. 70

tens	ones

b. 76

tens	ones

c. ____

tens	ones
4	9

d. ____

tens	ones
9	4

e. 65

tens	ones

f. 60

tens	ones

g. 90

tens	ones

h. ____

tens	ones
10	0

i. ____

tens	ones
8	3

j. ____

tens	ones
8	0

Lesson 3: Use the place value chart to record and name tens and ones within a two-digit number up to 100.

A STORY OF UNITS — Lesson 3 Fluency Template — 1•6

0   1   2   3

4   5   6   7

8   9   10   5

=   +   +   −

_____

numeral cards

Lesson 3: Use the place value chart to record and name tens and ones within a two-digit number up to 100.

A STORY OF UNITS Lesson 3 Template 1 1•6

1 0	2 0
3 0	4 0
5 0	6 0
7 0	8 0

Hide Zero cards, numeral side. Copy double-sided, and replace the cards from Module 4.

Lesson 3: Use the place value chart to record and name tens and ones within a two-digit number up to 100.

A STORY OF UNITS

Lesson 3 Template 1 1•6

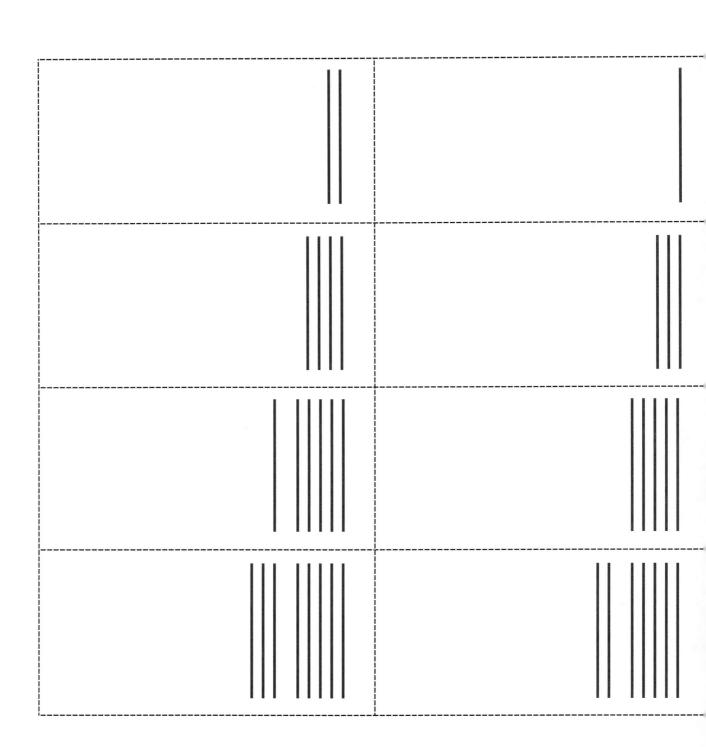

Hide Zero cards, quick tens side. Copy double-sided, and replace the cards from Module 4.

Lesson 3: Use the place value chart to record and name tens and ones within a two-digit number up to 100.

A STORY OF UNITS

Lesson 3 Template 1 1•6

9 0

1 0 0

Hide Zero cards, numeral side. Copy double-sided, and replace the cards from Module 4.

Lesson 3: Use the place value chart to record and name tens and ones within a two-digit number up to 100.

65

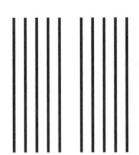

Hide Zero cards, quick tens side. Copy double-sided, and replace the cards from Module 4.

A STORY OF UNITS

Lesson 3 Template 2 1•6

tens	ones

tens	ones

place value chart

Lesson 3: Use the place value chart to record and name tens and ones within a two-digit number up to 100.

67

Lesson 4

Objective: Write and interpret two-digit numbers to 100 as addition sentences that combine tens and ones.

Suggested Lesson Structure

- ■ Application Problem   (5 minutes)
- ■ Fluency Practice   (17 minutes)
- ■ Concept Development   (28 minutes)
- ■ Student Debrief   (10 minutes)

  Total Time   **(60 minutes)**

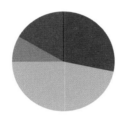

Application Problem (5 minutes)

Tamra has 14 goldfish. Darnel has 8 goldfish. How many fewer goldfish does Darnel have than Tamra?

Note: Today's Application Problem presents a *compare with difference unknown* problem type. Continue to ask students the following questions:

- Can you draw something?
- What can you draw?
- What does your drawing show you that can help answer the question?

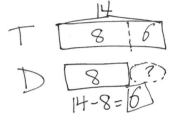

Darnel has 6 fewer goldfish

Fluency Practice (17 minutes)

- Grade 1 Core Fluency Sprint **1.OA.6**   (10 minutes)
- Digit Detective **1.NBT.2**   (4 minutes)
- Tens and Ones **1.NBT.4**   (3 minutes)

A STORY OF UNITS   Lesson 4 1•6

Grade 1 Core Fluency Sprint (10 minutes)

Materials: (S) Core Fluency Sprints (Lesson 3)

Note: Based on the needs of the class, select a Sprint from yesterday's materials. There are several possible options available.

1. Re-administer the Sprint from the day before.
2. Administer the next Sprint in the sequence.
3. Differentiate. Administer two different Sprints. Simply have one group do a counting activity on the back of the Sprint while the other Sprint is corrected.

Hopefully, the daily Sprints and Practice Sets are helping students to make good progress toward mastering the required Core Fluency for Grade 1. Support students who regularly finish fewer than half of the problems on a Sprint. Take note of the problem types that slow them down. Perhaps send the next day's Sprint home with them the night prior to administration. Awareness of a student's weak spots facilitates targeted support from within the learning community. For example, a volunteer can be charged with helping a certain student gain fluency with subtracting 3 from numbers within 10.

Digit Detective (4 minutes)

Materials: (T/S) Personal white board

Note: This activity reviews place value for two-digit numbers to 100, which was introduced in the previous lesson. Allow students to use their personal white boards to record the mystery numbers as needed.

Write a number on your personal white board, but do not show students.

- T: The digit in the tens place is 2. The digit in the ones place is 1. What's my number?
- S: 21.
- T: What's the value of the 2? (Pause, and then snap.)
- S: 20.
- T: What's the value of the 1? (Pause, and then snap.)
- S: 1.
- T: (Reveal the number.)

Repeat with the following suggested sequence: 12, 45, 54, 63, 87, 78, and 92. Alternate saying the number in the ones place first and saying the number in the tens place first. For the last minute, challenge students with adding or subtracting clues for mystery numbers between 40 and 99 as in the examples below.

- T: The digit in the tens place is 1 more than 3. (Pause.) The digit in the ones place is 10 less than 12. Say the number the Say Ten way.
- S: 4 tens 2.
- T: The digit in the ones place is equal to 5 + 3. The digit in the tens place is equal to 10 – 5. Say the number the Say Ten way.
- S: 5 tens 8.
- T: (Reveal the number.)

Lesson 4: Write and interpret two-digit numbers to 100 as addition sentences that combine tens and ones.

69

A STORY OF UNITS
Lesson 4 1•6

Tens and Ones (3 minutes)

Materials: (T) Rekenrek

Note: Reviewing this Module 4 fluency activity prepares students for today's lesson.

Practice decomposing numbers into tens and ones using the Rekenrek.

- T: (Show 16 on the Rekenrek.) How many tens do you see?
- S: 1.
- T: How many ones?
- S: 6.
- T: Say the number the Say Ten way.
- S: Ten 6.
- T: 1 ten plus 6 ones is …?
- S: 16.

Slide over the next row, and repeat the process for 26 and 36. Continue with the following suggested sequence within 40: 15, 25, 35, 17, 27, 37, 19, 29, and 39.

Concept Development (28 minutes)

Materials: (T) Chart paper with a place value chart, Hide Zero cards (Lesson 3 Template 1) (S) Personal white board, place value chart (Lesson 3 Template 2), numeral cards (Lesson 3 Fluency Template)

Gather students in the meeting area in a semicircle formation with their personal white boards.

- T: (Show 78 with Hide Zero cards.) When I pull apart these Hide Zero cards, what two numbers will you see?
- S: 70 and 8.
- T: (Pull apart the Hide Zero cards.) How many tens are in 70? Record the tens in your place value chart.
- S: 7 tens. (Write 7 in the tens place.)
- T: How many ones are here? (Show the 8 card.) Fill in the rest of your place value chart.
- S: 8 ones. (Write 8 in the ones place.)
- T: Say this number as tens and ones.
- S: 7 tens 8 ones.
- T: 7 tens and 8 ones is the same as …?
- S: 78.
- T: On your personal white board, make a number bond that shows the tens and the ones.
- S: (Break apart 78 into 70 and 8.)

$70 + 8 = 78$
$8 + 70 = 78$
$78 = 70 + 8$
$78 = 8 + 70$
8 more than 70 is 78.
70 more than 8 is 78.

70 | Lesson 4: Write and interpret two-digit numbers to 100 as addition sentences that combine tens and ones.

A STORY OF UNITS   Lesson 4 1•6

T: (Record the number bond on the chart.) Write as many addition sentences as you can that use your number bond.

Circulate and ensure that students are only using the three numbers from this bond: 78, 70, and 8. If students begin writing subtraction sentences, remind them of the directions. The teacher may choose to challenge some students to consider subtraction sentences, but these sentences are not addressed during the course of the lesson.

T: Give me a number sentence that matches this number bond. Start with the part that represents the tens. (Record on the chart as students answer.)
S: 70 + 8 = 78.
T: Start your number sentence with the ones. (Record on the chart.)
S: 8 + 70 = 78.
T: 78 is the same as…?
S: 70 plus 8. (Write 78 = 70 + 8 as students answer.)
T: This time, start with the ones. 78 is the same as…?
S: 8 plus 70. (Write 78 = 8 + 70 as students answer.)
T: Talk to your partner. What do you notice about the addends in all of these number sentences?
S: 70 is a bigger number than 8. → They always say how many tens and ones make up the total. → You can switch the addends around, and the total is still the same.
T: Let's make some *more than* statements. 8 more than 70 is…? Say the whole sentence.
S: 8 more than 70 is 78. (Record on the chart.)
T: 70 more than 8 is…? Say the whole sentence.
S: 70 more than 8 is 78. (Record on the chart.)

NOTES ON MULTIPLE MEANS OF EXPRESSION:

Students may need additional support with the language of "___ is the same as ___," "___ is ___ more than ___," etc. Insert a sentence frame into the personal white board, and allow the student to fill in the blanks. Pointing to each word and number as it is read can provide a bridge between the concrete and the abstract.

Repeat the process following the suggested sequence: 54, 62, 75, 57, 83, 91, and 100. Use different language to elicit a variety of answers for each number. For example, 54 is the same as____, 50 plus 4 is_____, 5 tens and 4 ones is____, 4 more than 50 is____, and 50 more than 4 is____.

For the remainder of time, have partners play Combine Tens and Ones. Leave the chart for 78 up on the board as a reference to support students.

NOTES ON MULTIPLE MEANS OF ENGAGEMENT:

When playing games with students, modeling how the game is played is very important. Oral instructions alone do not help everyone in the class understand how the game is played. Have two students demonstrate Partner A and Partner B roles so that all students see and hear the way the game is played.

- Prepare two decks of numeral cards 0 through 9 for each pair.
- Pick a card from the first deck. This number is placed in the tens place on the place value chart. For example, 7 is drawn and placed in the tens place.

Lesson 4: Write and interpret two-digit numbers to 100 as addition sentences that combine tens and ones.

71

A STORY OF UNITS   Lesson 4 1•6

- Pick a card from the second deck. This number is placed in the ones place on the place value chart.
  For example, 5 is drawn and placed in the ones place.
- Partners A and B make a number bond decomposing the number into tens and ones.
- Partner A writes two addition number sentences, such as those in the image from the previous page.
- Partner B writes a *more than* statement that combines tens and ones, such as those in the image on the previous page.
- Switch roles for the next pair of cards drawn.

70 + 5 = 75
5 + 70 = 75
75 = 70 + 5
75 = 5 + 70

70 more than 5 is 75.
5 more than 70 is 75.
75 is 5 more than 70.
75 is 70 more than 5.

Problem Set (10 minutes)

Students should do their personal best to complete the Problem Set within the allotted 10 minutes. For some classes, it may be appropriate to modify the assignment by specifying which problems they work on first. Some problems do not specify a method for solving. Students should solve these problems using the RDW approach used for Application Problems.

Student Debrief (10 minutes)

Lesson Objective: Write and interpret two-digit numbers to 100 as addition sentences that combine tens and ones.

The Student Debrief is intended to invite reflection and active processing of the total lesson experience.

Invite students to review their solutions for the Problem Set. They should check work by comparing answers with a partner before going over answers as a class. Look for misconceptions or misunderstandings that can be addressed in the Debrief. Guide students in a conversation to debrief the Problem Set and process the lesson.

Any combination of the questions below may be used to lead the discussion.

- For Problems 3 and 4, even though the totals use the same digits, the value of each answer is different. Explain why this is so.
- Look at Problem 10. How many tens make up 100? How can you express 100 as all ones?
- Look at Problem 1. If we unbundled one of the tens, how many tens and ones will we have?

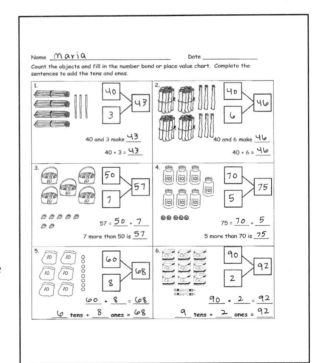

Lesson 4: Write and interpret two-digit numbers to 100 as addition sentences that combine tens and ones.

A STORY OF UNITS

Lesson 4 1•6

- Look at Problems 3, 4, and 5. What do you think are in the baskets? In the bottles? In the bags? What makes you think this?
- How did today's fluency activities connect with today's lesson?
- How did you solve the Application Problem? What other problems did this one remind you of?

Exit Ticket (3 minutes)

After the Student Debrief, instruct students to complete the Exit Ticket. A review of their work will help with assessing students' understanding of the concepts that were presented in today's lesson and planning more effectively for future lessons. The questions may be read aloud to the students.

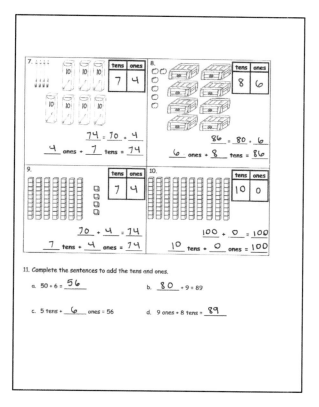

Lesson 4: Write and interpret two-digit numbers to 100 as addition sentences that combine tens and ones.

73

A STORY OF UNITS Lesson 4 Problem Set 1•6

Name _____ Date _____

Count the objects, and fill in the number bond or place value chart. Complete the sentences to add the tens and ones.

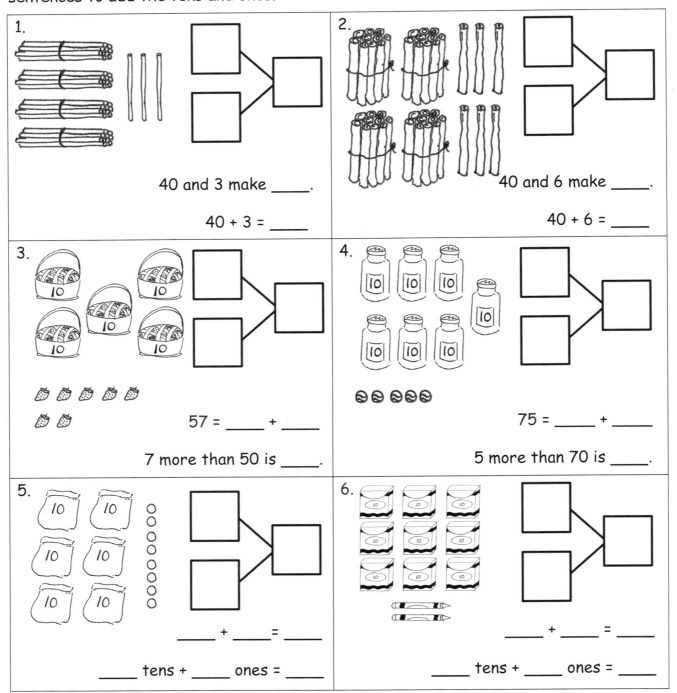

1. 40 and 3 make ____.
   40 + 3 = ____

2. 40 and 6 make ____.
   40 + 6 = ____

3. 57 = ____ + ____
   7 more than 50 is ____.

4. 75 = ____ + ____
   5 more than 70 is ____.

5. ____ + ____ = ____
   ____ tens + ____ ones = ____

6. ____ + ____ = ____
   ____ tens + ____ ones = ____

Lesson 4: Write and interpret two-digit numbers to 100 as addition sentences that combine tens and ones.

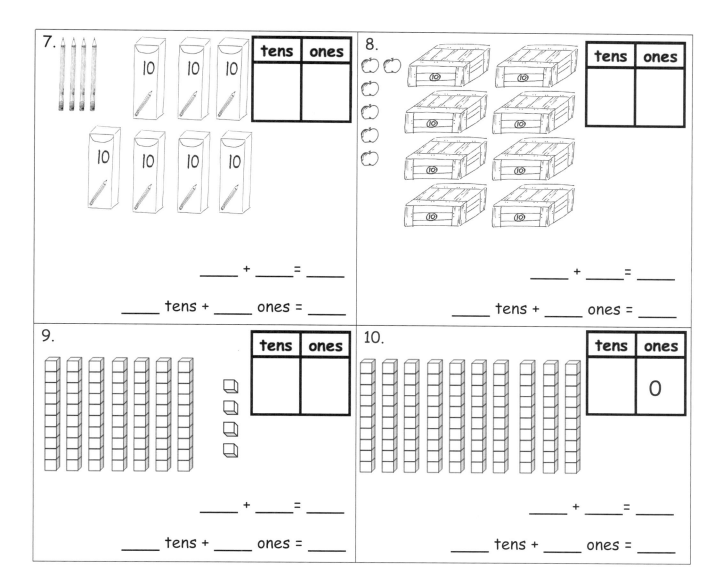

11. Complete the sentences to add the tens and ones.

   a. 50 + 6 = ____

   b. ____ + 9 = 89

   c. 5 tens + ____ ones = 56

   d. 9 ones + 8 tens = ____

A STORY OF UNITS   Lesson 4 Exit Ticket 1•6

Name _____ Date _____

1. Count the objects, and fill in the number bond or place value chart. Complete the sentences to add the tens and ones.

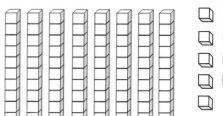 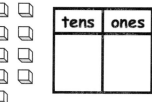

_____ + _____ = _____

____ tens + ____ ones = ____

2. Complete the sentences to add the tens and ones.

   a. 90 + 2 = ____

   b. 7 tens + ____ ones = 79

A STORY OF UNITS Lesson 4 Homework 1•6

Name _____ Date _____

Count the objects, and fill in the number bond or place value chart. Complete the sentences to add the tens and ones.

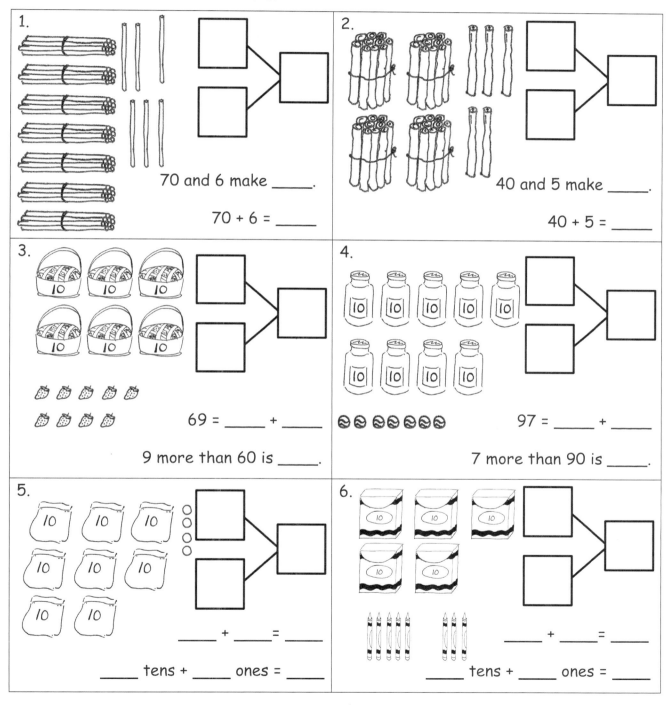

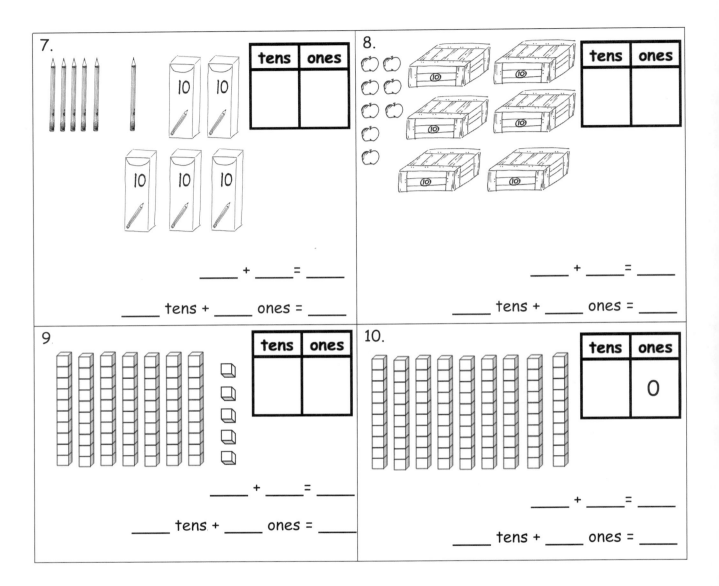

11. Complete the sentences to add the tens and ones.

   a. 80 + 6 = ____

   b. ____ + 7 = 57

   c. 9 tens + ____ ones = 95

   d. 4 ones + 8 tens = ____

Lesson 5

Objective: Identify 10 more, 10 less, 1 more, and 1 less than a two-digit number within 100.

Suggested Lesson Structure

- ■ Application Problem (5 minutes)
- ■ Fluency Practice (13 minutes)
- ■ Concept Development (32 minutes)
- ■ Student Debrief (10 minutes)

Total Time **(60 minutes)**

Application Problem (5 minutes)

Kiana has 6 fewer goldfish than Tamra. Tamra has 14 goldfish. How many goldfish does Kiana have?

Note: Today's Application Problem is the last in a series of three problems that use a related context. The three problems can be discussed together during the Student Debrief. As students share strategies and compare and contrast the problem stories, they gain a stronger sense of each particular problem type.

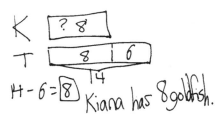

Fluency Practice (13 minutes)

- Core Fluency Differentiated Practice Sets **1.OA.6**   (5 minutes)
- Subtraction with Cards **1.OA.6**   (5 minutes)
- Coin Drop **1.NBT.5, 1.MD.3**   (3 minutes)

Core Fluency Differentiated Practice Sets (5 minutes)

Materials: (S) Core Fluency Practice Sets (Lesson 1)

Note: Give the appropriate Practice Set to each student. Students who completed all of the questions on their most recent Practice Set correctly should be given the next level of difficulty. All other students should try to improve their scores on their current levels.

Have students complete as many problems as they can in 90 seconds. Assign a counting pattern and start number for early finishers, or have them practice make ten addition or subtraction on the back of their papers. Collect and correct any Practice Sets completed within the allotted time.

| A STORY OF UNITS | Lesson 5 1•6 |

Subtraction with Cards (5 minutes)

Materials: (S) 1 pack of numeral cards 0–10 (Lesson 3 Fluency Template)

Note: This review activity targets the core subtraction fluency for Grade 1. As students play, closely monitor any students who have not performed well on the core Practice Sets and Sprints to see if they are able to be successful in this untimed, interactive game. Take advantage of any opportunity to highlight improvement.

- Students combine their digit cards and place them facedown between them.
- Each partner flips over two cards and subtracts the smaller number from the larger one.
- The partner with the smallest difference keeps the cards played by both players in that round.
- If the differences are equal, the cards are set aside, and the winner of the next round keeps the cards from both rounds.
- A player wins by having the most cards when the time is up.

Coin Drop (3 minutes)

Materials: (T) 4 dimes, 10 pennies, can

Note: In this activity, students practice adding and subtracting ones and tens within 40. This skill is expanded to numbers within 100 in today's lesson.

- T: (Hold up a penny.) Name my coin.
- S: A penny.
- T: How much is it worth?
- S: 1 cent.
- T: Listen carefully as I drop coins in my can. Count along in your minds.

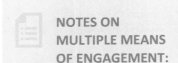

NOTES ON MULTIPLE MEANS OF ENGAGEMENT:

After playing Coin Drop with pennies and then dimes, mix pennies and dimes so that students have to add based on the changing value of the coin. Alternate between naming the coin and the value before dropping the coins. This challenges students and keeps them listening for what comes next.

Drop in some pennies, and ask how much money is in the can. Take out some pennies, and show them. Ask how much money is still in the can. Continue adding and subtracting pennies for a minute or so. Then, repeat the activity with dimes.

Concept Development (32 minutes)

Materials: (T) 2 pieces of chart paper with two pairs of place value charts as shown (S) Personal white board, place value chart (Lesson 3 Template 2)

Have students sit at their desks with all materials.

- T: Draw 62 using a quick ten drawing.
- S: (Draw 6 quick tens and 2 circles.)
- T: According to your picture, how many tens and ones are in 62?
- S: 6 tens and 2 ones.

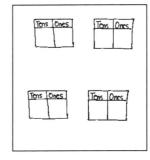

Lesson 5: Identify 10 more, 10 less, 1 more, and 1 less than a two-digit number within 100.

A STORY OF UNITS   Lesson 5 1•6

T: (Write 62 on the double place value chart template.)
T: Show me 1 more than 62.
S: (Draw 1 more circle.)
T: What is 1 more than 62? Say the whole sentence.
S: 1 more than 62 is 63. (Write 63 on the second place value chart.)
T: From 62 to 63, we added 1 more. (Draw an arrow from the first place value chart to the second, and write + 1 above the arrow.)
T: Look at the place value chart. Turn and explain to your partner about what did and did not change.
S: The tens didn't change. They both stayed as 6 tens because we only added 1 more. → The ones changed from 2 to 3 because we added 1 more. 3 is 1 more than 2. → To figure out 1 more, I just have to add 1 more to the ones place! (Note: In Problem 3 of the Problem Set, when dealing with 1 more than 89, the common misconception voiced by the last student is used as a talking point in the Debrief.)
T: Show me 62 with your drawing again.
S: (Show 62.)
T: (Write 62 on a new place value chart.) How can you show 10 more than 62? (Draw an arrow, and write + 10 above it.) Turn and talk to your partner.
S: Just draw 1 more quick ten!
T: Do that.
T: What is 10 more than 62? Say the whole sentence.
S: 10 more than 62 is 72.

MP.4

T: (Write 72 into the second place value chart.) Talk to your partner about what changes and what stays the same.
S: The tens changed this time from 6 tens to 7 tens because we added 10 more. → The ones didn't change because we just added 1 ten. → We could add 10 extra circles, but once you get 10, we make them into a quick ten, so why bother? We can add a ten quickly. → I just have to add 1 ten to the tens!
T: We added 10 more to 62 and now have 72.

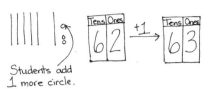

Students add 1 more circle.

NOTES ON MULTIPLE MEANS OF ENGAGEMENT:

Some students may not be able to imagine adding or subtracting a ten at this point. Support these students with all of the materials used in the lesson, and give them plenty of practice. Their path to abstract thinking may be a little longer than that of other students.

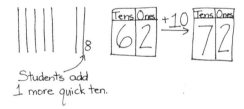

Students add 1 more quick ten.

NOTES ON MULTIPLE MEANS OF ENGAGEMENT:

Other students in the class may be able to visualize adding and subtracting ones and tens. Since these students have moved from concrete to abstract thinking, challenge them by giving problems adding or subtracting 2 ones or tens or 3 ones or tens.

Repeat the process using *1 less* and *10 less* with 87 as shown.

Lesson 5: Identify 10 more, 10 less, 1 more, and 1 less than a two-digit number within 100.

A STORY OF UNITS   **Lesson 5   1•6**

Then, follow the suggested sequence:

- 1 less/10 less than 77
- 1 more/1 less than 90

Problem Set (10 minutes)

Students should do their personal best to complete the Problem Set within the allotted 10 minutes. For some classes, it may be appropriate to modify the assignment by specifying which problems they work on first. Some problems do not specify a method for solving. Students should solve these problems using the RDW approach used for Application Problems.

Student Debrief (10 minutes)

Lesson Objective: Identify 10 more, 10 less, 1 more, and 1 less than a two-digit number within 100.

The Student Debrief is intended to invite reflection and active processing of the total lesson experience.

Invite students to review their solutions for the Problem Set. They should check work by comparing answers with a partner before going over answers as a class. Look for misconceptions or misunderstandings that can be addressed in the Debrief. Guide students in a conversation to debrief the Problem Set and process the lesson.

Any combination of the questions below may be used to lead the discussion.

- I say, "When I find 1 more, only the ones digit changes." I'm wrong! Which problem shows that I'm wrong? When am I correct?
- I say, "When I find 1 less, only the ones digit changes." I'm wrong! Which problem shows that I'm wrong again?
- How can you use the place value chart to help you count by ones? By tens?
- How did our fluency activity of Coin Drop relate to today's lesson?

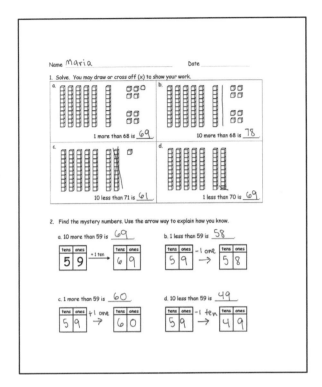

Lesson 5: Identify 10 more, 10 less, 1 more, and 1 less than a two-digit number within 100.

- Look at your Application Problem. How is it similar, and how is it different from other Application Problems you have solved? Share your strategy for beginning to solve the problem.

Exit Ticket (3 minutes)

After the Student Debrief, instruct students to complete the Exit Ticket. A review of their work will help with assessing students' understanding of the concepts that were presented in today's lesson and planning more effectively for future lessons. The questions may be read aloud to the students.

Lesson 5: Identify 10 more, 10 less, 1 more, and 1 less than a two-digit number within 100.

A STORY OF UNITS

Lesson 5 Problem Set 1•6

Name _____ Date _____

1. Solve. You may draw or cross off (x) to show your work.

a.

1 more than 68 is _____.

b.

10 more than 68 is _____.

c.

10 less than 71 is _____.

d.

1 less than 70 is _____.

2. Find the mystery numbers. Use the arrow way to explain how you know.

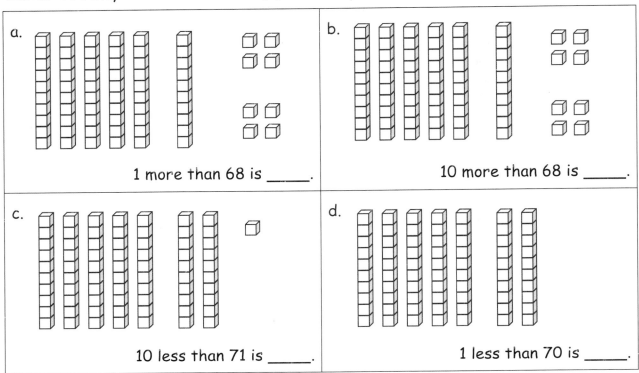

a. 10 more than 59 is _____.

b. 1 less than 59 is _____.

c. 1 more than 59 is _____.

d. 10 less than 59 is _____.

Lesson 5: Identify 10 more, 10 less, 1 more, and 1 less than a two-digit number within 100.

A STORY OF UNITS   Lesson 5 Problem Set 1•6

3. Write the number that is **1 more**.

   a. 10, _____
   b. 70, _____
   c. 76, _____
   d. 79, _____
   e. 99, _____

4. Write the number that is **10 more**.

   a. 10, _____
   b. 60, _____
   c. 61, _____
   d. 78, _____
   e. 90, _____

5. Write the number that is **1 less**.

   a. 12, _____
   b. 52, _____
   c. 51, _____
   d. 80, _____
   e. 100, _____

6. Write the number that is **10 less**.

   a. 20, _____
   b. 60, _____
   c. 74, _____
   d. 81, _____
   e. 100, _____

7. Fill in the missing numbers in each sequence.

   a. 40, 41, 42, _____
   b. 89, 88, 87, _____
   c. 72, 71, _____, 69
   d. 63, _____, 65, 66
   e. 40, 50, 60, _____
   f. 80, 70, 60, _____
   g. 55, 65, _____, 85
   h. 99, 89, _____, 69
   i. _____, 99, 98, 97
   j. _____, 77, _____, 57

Lesson 5:   Identify 10 more, 10 less, 1 more, and 1 less than a two-digit number within 100.

A STORY OF UNITS   Lesson 5 Exit Ticket 1•6

Name _____   Date _____

1. Find the mystery numbers. Use the arrow way to show how you know.

   a. 1 less than 69 is _____.   b. 10 more than 69 is _____.

tens	ones		tens	ones		tens	ones		tens	ones
------	------		------	------		------	------		------	------

2. Write the number that is **1 more**.	3. Write the number that is **10 more**.
a. 40, _____	a. 50, _____
b. 86, _____	b. 62, _____
c. 89, _____	c. 90, _____
4. Write the number that is **1 less**.	5. Write the number that is **10 less**.
a. 75, _____	a. 80, _____
b. 70, _____	b. 99, _____
c. 100, _____	c. 100, _____

Lesson 5: Identify 10 more, 10 less, 1 more, and 1 less than a two-digit number within 100.

A STORY OF UNITS

Lesson 5 Homework 1•6

Name _____ Date _____

1. Solve. You may draw or cross off (x) to show your work.

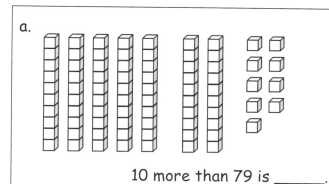

a. 10 more than 79 is _____.

b. 10 less than 81 is _____.

c. 1 more than 79 is _____.

d. 1 less than 80 is _____.

2. Find the mystery numbers. You may make a drawing to help solve, if needed.

a. 10 more than 75 is _____.

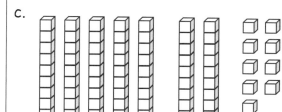

b. 1 more than 75 is _____.

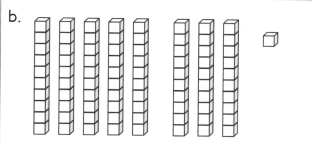

c. 10 less than 88 is _____.

d. 1 less than 88 is _____.

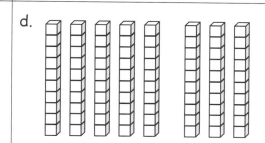

Lesson 5: Identify 10 more, 10 less, 1 more, and 1 less than a two-digit number within 100.

A STORY OF UNITS   **Lesson 5 Homework 1•6**

3. Write the number that is **1 more**.
   - a. 40, _____
   - b. 50, _____
   - c. 65, _____
   - d. 69, _____
   - e. 99, _____

4. Write the number that is **10 more**.
   - a. 60, _____
   - b. 70, _____
   - c. 77, _____
   - d. 89, _____
   - e. 90, _____

5. Write the number that is **1 less**.
   - a. 53, _____
   - b. 73, _____
   - c. 71, _____
   - d. 80, _____
   - e. 100, _____

6. Write the number that is **10 less**.
   - a. 50, _____
   - b. 60, _____
   - c. 84, _____
   - d. 91, _____
   - e. 100, _____

7. Fill in the missing numbers in each sequence.

   - a. 50, 51, 52, _____
   - b. 79, 78, 77, _____
   - c. 62, 61, _____, 59
   - d. 83, _____, 85, 86
   - e. 60, 70, 80, _____
   - f. 100, 90, 80, _____
   - g. 57, 67, _____, 87
   - h. 89, 79, _____, 59
   - i. _____, 99, 98, 97
   - j. _____, 84, _____, 64

A STORY OF UNITS

Lesson 6 1•6

Lesson 6

Objective: Use the symbols >, =, and < to compare quantities and numerals to 100.

Suggested Lesson Structure

- Application Problem   (5 minutes)
- Fluency Practice   (13 minutes)
- Concept Development   (32 minutes)
- Student Debrief   (10 minutes)

Total Time   **(60 minutes)**

Application Problem (5 minutes)

Nikil has 12 toy cars. Willie has 4 toy cars. When Nikil and Willie play, how many toy cars do they have?

Note: Today, the very simple *put together with result unknown* problem type is revisited. Please use this to highlight that students might use either a double or single tape to model as is pictured to the right.

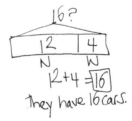 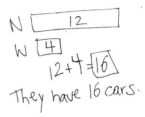

Fluency Practice (13 minutes)

- Core Fluency Differentiated Practice Sets **1.OA.6**   (5 minutes)
- Coin Drop **1.NBT.5, 1.MD.3**   (3 minutes)
- True or False Number Sentences **1.OA.6, 1.OA.7**   (5 minutes)

Core Fluency Differentiated Practice Sets (5 minutes)

Materials: (S) Core Fluency Practice Sets (Lesson 1)

Note: Give the appropriate Practice Set to each student. Help students become aware of their improvement by asking them to quickly stand if they tried a new level or improved their score from the previous day.

Students complete as many problems as they can in 90 seconds. Assign a counting pattern and start number for early finishers,

NOTES ON MULTIPLE MEANS OF ENGAGEMENT:

For students who are still on Practice Set A, a Practice Set (completed aloud) that is privately administered may help them be more successful. The pencil and paper can hold back some students who may have trouble with their fine motor skills.

Lesson 6:   Use the symbols >, =, and < to compare quantities and numerals to 100.

89

or have them practice make ten addition or subtraction on the back of their papers. Collect and correct any Practice Sets completed within the allotted time.

Coin Drop (3 minutes)

Materials: (T) 10 dimes, 10 pennies, can

Note: This activity reviews yesterday's lesson, in which students learned to add and subtract ones and tens within 100.

Today, start with 5 dimes in the can. Drop a penny or a dime into the can, asking students the total after each drop of one coin. Ask them to say, "1 cent more is 51 cents," or "10 cents more is 60 cents." For today, perhaps limit it to 1 more and 10 more.

True or False Number Sentences (5 minutes)

Materials: (T/S) Personal white board

Note: This activity provides practice with Grade 1's core fluency, while reviewing the inequality symbols that were presented in Module 4 Topic B.

Review the symbols =, >, and <. Write true and false number sentences using the symbols. On the signal, students say whether the number sentence is true or false. Then, choose a student who answered correctly to prove it.

- T (Write 5 = 7.) Is this number sentence true or false? (Pause, and then snap.)
- S: False.
- T: Why? Student A.
- S: 5 is less than 7.
- T: (Write 8 = 6 + 2.) True or false? (Pause, and then snap.)
- S: True.
- T: Why? Student B.
- S: 6 + 2 is 8, and 8 *is the same as* 8.
- T: (Write 8 = 8 underneath 8 = 6 + 2.)

Continue with the following suggested sequence. Be sure to space the number sentences so students can easily see the two expressions, and provide time for students to solve on their personal white boards as needed. Before completing the > and < columns (see below), write the symbol in the middle of the board, and review its meaning.

  a. 6 = 8 − 2   e. 5 > 6   i. 8 < 9
  b. 3 = 8 − 5   f. 7 > 4   j. 6 < 5
  c. 5 + 1 = 4 + 1   g. 8 > 7   k. 6 < 3 + 3
  d. 5 + 1 = 4 + 2   h. 6 > 9   l. 5 + 2 < 2 + 5

A STORY OF UNITS

Lesson 6 1•6

Concept Development (32 minutes)

Materials: (T) Chart paper, comparison cards (Template), tape (S) Personal white board, place value chart (Lesson 3 Template 2), comparison cards (Template)

Gather students in the meeting area with their materials.

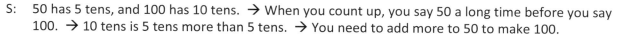

T: (On chart paper, write 100 and 50 in the place value charts with room between them to insert a comparison card.) Which number is greater?
S: 100.
T: How do you know?
S: 50 has 5 tens, and 100 has 10 tens. → When you count up, you say 50 a long time before you say 100. → 10 tens is 5 tens more than 5 tens. → You need to add more to 50 to make 100.
T: (Show < and > cards.) Which symbol should I use?
S: Greater than! → The one on the right.
T: (Tape the > symbol between the two place value charts.) What are some of the ways you help yourself remember that *this* (point to >) is the greater than symbol?
S: Pretend the open side is a hungry alligator's mouth that eats the bigger number. → The side with two endpoints is near the greater number. The side with 1 endpoint is near the smaller one.
T: (Tape the other two symbol cards to the chart paper.) What is the name of this symbol?
S: Less than!
T: This one?
S: Equal to!
T: Choose the symbol you think I should use to compare the two numbers I write. Wait for the snap.
T: (Write 60 and 90. Pause before giving the signal. Add the < symbol between 60 and 90.) Let's read our math sentence together.
S/T: 60 is less than 90.

Repeat the above process with the following suggested sequence of numbers: 59 and 52, 80 and 70, 49 and 94, 7 tens and 6 tens 8 ones, 78 ones and 8 tens, 67 ones and 6 tens, 7 tens and 6 tens 10 ones, 10 tens and 90, and 8 tens 2 ones and 7 tens and 15 ones.

If students could use more practice, invite them to play Compare It! with a partner.

- Each partner writes a number from 0 to 100 on her white board, without showing her partner.
- When both are ready, they show their boards.
- For the first round, Partner A uses the cards to put the symbol between the boards.
- Partner B reads the true number sentence that was made. (Demonstrate with the number sentence on the board.)

NOTES ON MULTIPLE MEANS OF ENGAGEMENT:

As students work on the sequence of numbers, be sure they are reading the math sentence out loud once they choose the symbol to compare the numbers. Being able to read the sentence properly demonstrates they have mastered the difference between the symbols.

Lesson 6: Use the symbols >, =, and < to compare quantities and numerals to 100.

A STORY OF UNITS

Lesson 6 1•6

At the end of the first round, have partners use Partner B's cards. Alternate for each round until the students have played for four minutes. During that time, circulate and notice which students are successful and which students may need more support. Encourage students to make the game more challenging by varying how they represent the number, using quick tens, place value charts, and writing the numbers as tens and ones.

Problem Set (10 minutes)

Students should do their personal best to complete the Problem Set within the allotted 10 minutes. For some classes, it may be appropriate to modify the assignment by specifying which problems they work on first. Some problems do not specify a method for solving. Students should solve these problems using the RDW approach used for Application Problems.

Student Debrief (10 minutes)

Lesson Objective: Use the symbols >, =, and < to compare quantities and numerals to 100.

The Student Debrief is intended to invite reflection and active processing of the total lesson experience.

Invite students to review their solutions for the Problem Set. They should check work by comparing answers with a partner before going over answers as a class. Look for misconceptions or misunderstandings that can be addressed in the Debrief. Guide students in a conversation to debrief the Problem Set and process the lesson.

Any combination of the questions below may be used to lead the discussion.

- Look at Problem 1(g). How did you solve this problem? Explain your thinking.
- Which problem was the trickiest in the Problem Set to compare? What made it tricky, and how did you or your partner solve it? If you were going to give a friend advice on how to solve these kinds of tricky comparisons, what would you suggest to him?
- Share a comparison problem that you and your partner created during the Compare It! activity.

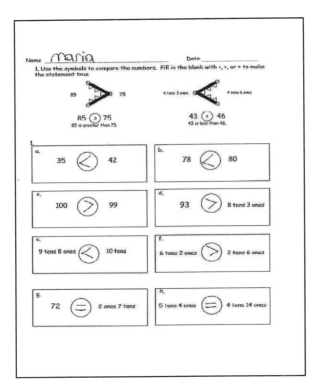

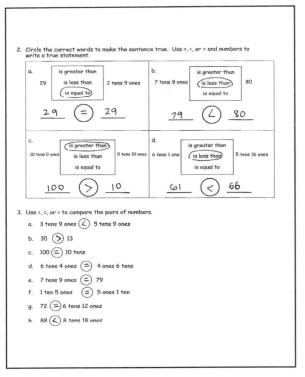

Lesson 6: Use the symbols >, =, and < to compare quantities and numerals to 100.

- With your partner, share how you remember the meaning of each symbol.
- How did today's Fluency Practice help you with our lesson? Explain your thinking.
- Look at your Application Problem. Share your drawing and your solution. How did your drawing help you solve the problem? How is your drawing similar to or different from your partner's drawing?

Exit Ticket (3 minutes)

After the Student Debrief, instruct students to complete the Exit Ticket. A review of their work will help with assessing students' understanding of the concepts that were presented in today's lesson and planning more effectively for future lessons. The questions may be read aloud to the students.

Lesson 6: Use the symbols >, =, and < to compare quantities and numerals to 100.

A STORY OF UNITS　　　　　　　　　　　　　　　　　　　　Lesson 6 Problem Set　1•6

Name _____　　Date _____

1. Use the symbols to compare the numbers. Fill in the blank with <, >, or = to make the statement true.

85 > 75　　　　　　4 tens 3 ones < 4 tens 6 ones

85 ◯(>) 75　　　　　　　　43 ◯(<) 46
85 is greater than 75.　　　　　43 is less than 46.

a. 35 ◯ 42

b. 78 ◯ 80

c. 100 ◯ 99

d. 93 ◯ 8 tens 3 ones

e. 9 tens 8 ones ◯ 10 tens

f. 6 tens 2 ones ◯ 2 tens 6 ones

g. 72 ◯ 2 ones 7 tens

h. 5 tens 4 ones ◯ 4 tens 14 ones

94　　Lesson 6:　Use the symbols >, =, and < to compare quantities and numerals to 100.

2. Circle the correct words to make the sentence true. Use >, <, or = and numbers to write a true statement.

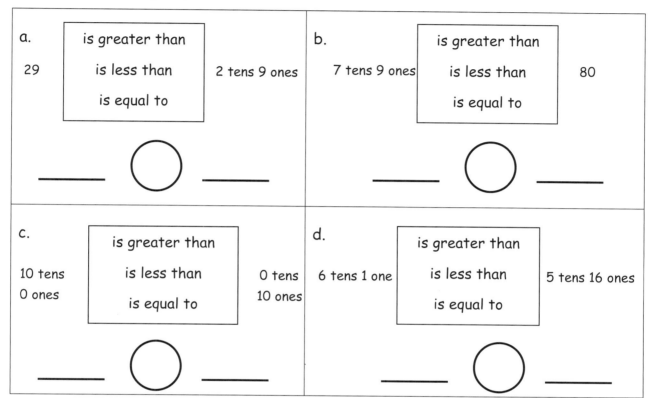

3. Use <, =, or > to compare the pairs of numbers.

   a. 3 tens 9 ones ◯ 5 tens 9 ones
   b. 30 ◯ 13
   c. 100 ◯ 10 tens
   d. 6 tens 4 ones ◯ 4 ones 6 tens
   e. 7 tens 9 ones ◯ 79
   f. 1 ten 5 ones ◯ 5 ones 1 ten
   g. 72 ◯ 6 tens 12 ones
   h. 88 ◯ 8 tens 18 ones

A STORY OF UNITS

Lesson 6 Exit Ticket 1•6

Name _____ Date _____

Circle the correct words to make the sentence true. Use >, <, or = and numbers to write a true statement.

a. 36 [is greater than / is less than / is equal to] 6 tens 3 ones

____ ◯ ____

b. 90 [is greater than / is less than / is equal to] 8 tens 9 ones

____ ◯ ____

c. 52 [is greater than / is less than / is equal to] 5 tens 2 ones

____ ◯ ____

d. 4 tens 2 ones [is greater than / is less than / is equal to] 3 tens 14 ones

____ ◯ ____

Name _____ Date _____

1. Use the symbols to compare the numbers. Fill in the blank with <, >, or = to make the statement true.

62 > 57
62 is greater than 57.

56 < 59
56 is less than 59.

a. 43 ◯ 35

b. 60 ◯ 86

c. 10 tens ◯ 99

d. 5 tens 4 ones ◯ 54

e. 7 tens 9 ones ◯ 9 tens 7 ones

f. 1 ten 3 ones ◯ 31

g. 3 tens 0 ones ◯ 2 tens 10 ones

h. 3 tens 5 ones ◯ 2 tens 17 ones

A STORY OF UNITS　　　　　　　　　　　　　　　　　　　　　　　Lesson 6 Homework　1•6

2. Fill in the correct words from the box to make the sentence true. Use >, <, or = and numbers to write a true statement.

| is greater than | is less than | is equal to |

a.　　　42 _____ 1 ten 2 ones

____ ◯ ____

b.　6 tens 7 ones _____ 5 tens 17 ones

____ ◯ ____

c.　　　37 _____ 73

____ ◯ ____

d.　2 tens 14 ones _____ 4 ones 2 tens

____ ◯ ____

e.　9 ones 5 tens _____ 9 tens 5 ones

____ ◯ ____

Lesson 6: Use the symbols >, =, and < to compare quantities and numerals to 100.

A STORY OF UNITS

Lesson 6 Template 1•6

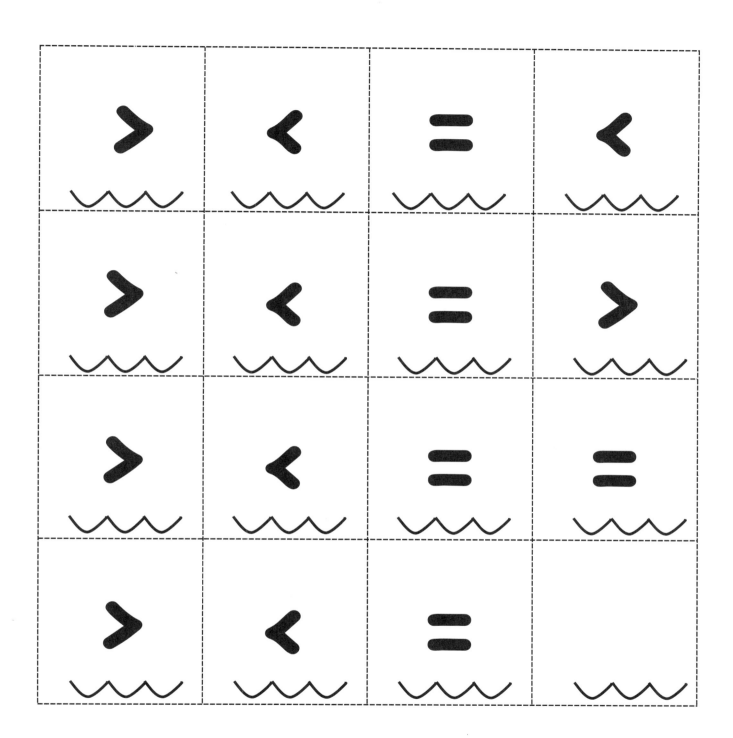

comparison cards, page 1. Print double-sided on cardstock. Distribute each of the three cards to students.

Lesson 6: Use the symbols >, =, and < to compare quantities and numerals to 100.

A STORY OF UNITS

Lesson 6 Template 1•6

less than	equal to	less than	greater than
greater than	equal to	less than	greater than
equal to	equal to	less than	greater than
	equal to	less than	greater than

comparison cards, page 2. Print double-sided on cardstock. Distribute each of the three cards to students.

Lesson 6: Use the symbols >, =, and < to compare quantities and numerals to 100.

A STORY OF UNITS

Lesson 7 1•6

Lesson 7

Objective: Count and write numbers to 120. Use Hide Zero cards to relate numbers 0 to 20 to 100 to 120.

Suggested Lesson Structure

- Application Problem (5 minutes)
- Fluency Practice (15 minutes)
- Concept Development (30 minutes)
- Student Debrief (10 minutes)

Total Time **(60 minutes)**

Application Problem (5 minutes)

Shanika has 6 roses and 7 tulips in a vase. Maria has 4 roses and 8 tulips in a vase. Who has more flowers? How many more flowers does she have?

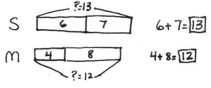

Note: Today's problem embeds an opportunity for comparison. Students continue to practice adding across ten, which supports their work in Topic C.

Fluency Practice (15 minutes)

- Grade 1 Core Fluency Sprint **1.OA.6** (10 minutes)
- True or False Number Sentences **1.OA.6, 1.OA.7** (5 minutes)

Grade 1 Core Fluency Sprint (10 minutes)

Materials: (S) Core Fluency Sprints (Lesson 3)

Note: Choose an appropriate Sprint based on the needs of the class. As students work, pay attention to their strategies and the number of problems they answer. Today, between Sides A and B of the Sprint, practice counting the Say Ten way up and down from 67 to 77.

Core Fluency Sprint List:

- Core Addition Sprint 1 (targeting core addition and missing addends)
- Core Addition Sprint 2 (targeting the most challenging addition within 10 and beyond 10)
- Core Subtraction Sprint (targeting core subtraction)

Lesson 7: Count and write numbers to 120. Use Hide Zero cards to relate numbers 0 to 20 to 100 to 120.

101

A STORY OF UNITS   Lesson 7 1•6

- Core Fluency Sprint: Totals of 5, 6, and 7 (developing understanding of the relationship between addition and subtraction)
- Core Fluency Sprint: Totals of 8, 9, and 10 (developing understanding of the relationship between addition and subtraction)

True or False Number Sentences (5 minutes)

Materials: (T/S) Personal white board

Note: This activity reviews Lesson 6.

Review the symbols =, >, and <. Write true and false number sentences using the symbols. Signal, and then wait for students to say whether the number sentence is true or false. Choose a student who answered correctly to prove it.

Use the first two columns as the suggested sequence. At each checkpoint, decide whether students are ready for the next column or if they should continue with similar problem types. The third column is provided as a possible opportunity for a few students who would really enjoy a challenge.

a. 5 > 4
b. 50 > 40
c. 45 > 54
d. 15 < 41
Checkpoint.

e. 30 + 5 = 35
f. 53 = 5 + 30
g. 73 < 7 tens 3 ones
h. 94 > 9 ones 3 tens
Checkpoint.

i. 9 + 8 = 10 + 7
j. 15 + 10 = 25 − 10
k. 14 − 7 > 9
l. 80 < 79 + 1

Concept Development (30 minutes)

Materials: (T) Vertical counting sequence (Template), Hide Zero cards (Lesson 3 Template 1) (S) Hide Zero cards (optional)

Have students sit at their desks at the start of the lesson. If students are using Hide Zero cards, distribute cards up to 9 tens. Hold students' 10 tens card until later in the lesson. The 11 tens and 12 tens cards are not needed for today's lesson.

- T: (Project the vertical counting sequence template, preferably on an interactive board or easel paper.) This chart shows numbers from 1 through 77. Can you help me write more numbers until we fill up all of the empty spaces?
- S: (Write the numbers on the chart as students count.) 78, 79, 80, 81, 82, ..., 100.
- T: We have more spaces on the chart. Who knows what number comes after 100?
- S: One hundred one.
- T: Yes. One hundred one (101), one hundred two (102), ... 120. (Be sure to read the number without saying *and* between one hundred and the ones place unit.)
- T: These last two columns look a little like other columns on the chart. Does anyone see what I see?
- S: The first two columns have most of the same digits.

102   Lesson 7:   Count and write numbers to 120. Use Hide Zero cards to relate numbers 0 to 20 to 100 to 120.

A STORY OF UNITS   **Lesson 7 1•6**

T: Let's look more closely at these columns. (If using an interactive board, highlight numbers 1 through 20 and numbers 101 through 120.) Talk with a partner about what you notice. (Circulate and listen as students discuss.)

S: I notice that there is a 1, 2, 3, 4, ... all the way to 20 at the beginning of this chart and at the end of this chart. → The pattern goes to 100 and starts over again, but you can't forget to include 100 each time as you say the new numbers. → Once you get to 100, the numbers start over again, only this time you say 100 first. So instead of 1, 2, 3, 4, it's 101, 102, 103, and 104.

T: Let's try this again with Hide Zero cards and see if we can tell what's happening. I'm going to give you a new Hide Zero card. This one has 10 tens. (Distribute 10 tens card.)

T: When we get to 100, the next number is…?

S: 101.

T: Point to the ones place on your hundred card.

S: (Students point.)

T: Place one on top of your 100 card in the ones place. What number did you make?

S: One hundred one!

T: Yes! It looks like a zero sandwich with the ones as the bread. What number is that again?

S: One hundred one!

T: Now, let's add another one. 100, 101, …?

S: 102.

T: Which card did you need to show 102?

S: Our 100 card and a 2.

NOTES ON MULTIPLE MEANS OF ENGAGEMENT:

Challenge students who seem confident with this skill to start at 120 and complete their chart by counting back.

NOTES ON MULTIPLE MEANS OF ENGAGEMENT:

If students have trouble completing the Problem Set because they do not know their numbers to 120, give them a vertical counting sequence to reference as they complete the problems.

MP.4

Repeat this process until the class reaches 120. Then, as a class, count down either verbally as the teacher points to the numbers on the chart displayed or as students create the numbers with their Hide Zero cards, until reaching 88.

Problem Set (10 minutes)

Students should do their personal best to complete the Problem Set within the allotted 10 minutes. For some classes, it may be appropriate to modify the assignment by specifying which problems they work on first. Some problems do not specify a method for solving. Students should solve these problems using the RDW approach used for Application Problems.

   Lesson 7:   Count and write numbers to 120. Use Hide Zero cards to relate numbers 0 to 20 to 100 to 120.   103

A STORY OF UNITS

Lesson 7 1•6

Student Debrief (10 minutes)

Lesson Objective: Count and write numbers to 120. Use Hide Zero cards to relate numbers 0 to 20 to 100 to 120.

The Student Debrief is intended to invite reflection and active processing of the total lesson experience.

Invite students to review their solutions for the Problem Set. They should check work by comparing answers with a partner before going over answers as a class. Look for misconceptions or misunderstandings that can be addressed in the Debrief. Guide students in a conversation to debrief the Problem Set and process the lesson.

Any combination of the questions below may be used to lead the discussion.

- Look at Problem 1. What are some patterns you notice in the chart?
- Look at Problem 4. Which sequences were the quickest for you to solve? Why? Which sequences were trickier? On your personal white board, create a really tricky problem for your partner. What did you do to make it tricky to solve? What strategies might you use to solve it correctly?
- Share the progress you have made with your work with Sprints. Tell us about the math accomplishments you are proud of.
- Look at your Application Problem. Share your strategies for solving the problem.

Exit Ticket (3 minutes)

After the Student Debrief, instruct students to complete the Exit Ticket. A review of their work will help with assessing students' understanding of the concepts that were presented in today's lesson and planning more effectively for future lessons. The questions may be read aloud to the students.

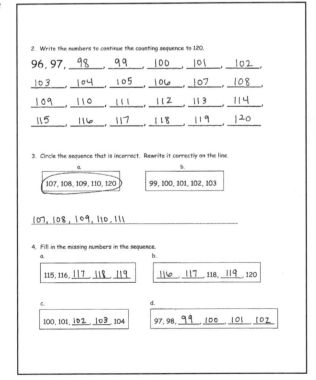

104   Lesson 7:   Count and write numbers to 120. Use Hide Zero cards to relate numbers 0 to 20 to 100 to 120.

A STORY OF UNITS

Lesson 7 Problem Set 1•6

Name _____ Date _____

1. Fill in the missing numbers in the chart up to 120.

a.	b.	c.	d.	e.
71	81	91		111
	82		102	
73	83	93		113
	84	94	104	114
76	86	96	106	116
77	87	97		117
79	89	99	109	119
80		100	110	

Lesson 7: Count and write numbers to 120. Use Hide Zero cards to relate numbers 0 to 20 to 100 to 120.

105

A STORY OF UNITS

Lesson 7 Problem Set 1•6

2. Write the numbers to continue the counting sequence to 120.

96, 97, _____, _____, _____, _____, _____,

_____, _____, _____, _____, _____, _____,

_____, _____, _____, _____, _____, _____,

_____, _____, _____, _____, _____, _____

3. Circle the sequence that is incorrect. Rewrite it correctly on the line.

a.

107, 108, 109, 110, 120

b.

99, 100, 101, 102, 103

_____

4. Fill in the missing numbers in the sequence.

a.

115, 116, _____, _____, _____

b.

_____, _____, 118, _____, 120

c.

100, 101, _____, _____, 104

d.

97, 98, _____, _____, _____, _____

A STORY OF UNITS

Lesson 7 Exit Ticket 1•6

Name _____ Date _____

1. Complete the chart by filling in the missing numbers.

   a.
88
90

   b.
99

   c.
108

   d.
119

2. Fill in the missing numbers to continue the counting sequence.

   a.
   117, _____, 119, _____

   b.
   108, 109, _____, _____, _____

Lesson 7: Count and write numbers to 120. Use Hide Zero cards to relate numbers 0 to 20 to 100 to 120.

A STORY OF UNITS

Lesson 7 Homework 1•6

Name _____ Date _____

1. Fill in the missing numbers in the chart up to 120.

a.	b.	c.	d.	e.
71		91		111
	82		102	
		93		
74				114
	85		105	
		96		116
	87			
			108	
79		99		119
80	90		110	

A STORY OF UNITS　　　　　　　　　　　　　　　　　　　　Lesson 7 Homework 1•6

2. Write the numbers to continue the counting sequence to 120.

   99, _____, 101, _____, _____, _____, _____, _____, _____,

   _____, _____, _____, _____, _____, _____, _____,

   _____, _____, _____, _____, _____, _____

3. Circle the sequence that is incorrect. Rewrite it correctly on the line.

   a.

   | 116, 117, 118, 119, 120 |

   b.

   | 96, 97, 98, 99, 100, 110 |

   _____

4. Fill in the missing numbers in the sequence.

   a.

   | 113, 114, _____, _____, _____ |

   b.

   | _____, _____, _____, 120 |

   c.

   | 102, _____, _____, _____ |

   d.

   | 88, 89, _____, _____, _____, _____, |

Lesson 7: Count and write numbers to 120. Use Hide Zero cards to relate numbers 0 to 20 to 100 to 120.

1	11	21	31	41	51	61	71	81	91	101	111
2	12	22	32	42	52	62	72	82	92	102	112
3	13	23	33	43	53	63	73	83	93	103	113
4	14	24	34	44	54	64	74	84	94	104	114
5	15	25	35	45	55	65	75	85	95	105	115
6	16	26	36	46	56	66	76	86	96	106	116
7	17	27	37	47	57	67	77	87	97	107	117
8	18	28	38	48	58	68	78	88	98	108	118
9	19	29	39	49	59	69	79	89	99	109	119
10	20	30	40	50	60	70	80	90	100	110	120

vertical counting sequence

Lesson 7: Count and write numbers to 120. Use Hide Zero cards to relate numbers 0 to 20 to 100 to 120.

| A STORY OF UNITS | Lesson 8 1•6 |

Lesson 8

Objective: Count to 120 in unit form using only tens and ones. Represent numbers to 120 as tens and ones on the place value chart.

Suggested Lesson Structure

■ Application Problem   (5 minutes)
■ Fluency Practice   (14 minutes)
■ Concept Development   (31 minutes)
■ Student Debrief   (10 minutes)
  Total Time   **(60 minutes)**

Application Problem (5 minutes)

Lee found 15 sparkly rocks. Kim found 8 sparkly rocks. How many more sparkly rocks did Lee find than Kim?

Note: Today's Application Problem is a *compare with difference unknown* problem type. For students who are successful with solving this problem when the term *more* is used, consider adjusting the question to ask how many *fewer* sparkly rocks Kim found. By asking both questions, the teacher can help students recognize that the same solution sentence can be used with either question.

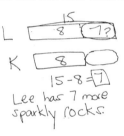

Fluency Practice (14 minutes)

- Grade 1 Core Fluency Sprint **1.OA.6**   (10 minutes)
- 1 More, 1 Less, 10 More, 10 Less **1.OA.5, 1.NBT.5**   (4 minutes)

Grade 1 Core Fluency Sprint (10 minutes)

Materials: (S) Core Fluency Sprints (Lesson 3)

Note: Based on the needs of the class, select a Sprint from yesterday's materials. There are several possible options available.

- Re-administer the Sprint from the day before.
- Administer the next Sprint in the sequence.
- Differentiate. Administer two different Sprints. Simply have one group do a counting activity on the back of the Sprint while the other Sprint is corrected.

Today, between Sides A and B of the Sprint, practice counting the Say Ten way from 88 to 99 and back.

 Lesson 8:   Count to 120 in unit form using only tens and ones. Represent numbers to 120 as tens and ones on the place value chart.   111

A STORY OF UNITS

Lesson 8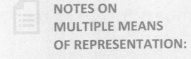

1 More, 1 Less, 10 More, 10 Less (4 minutes)

Materials: (T) Vertical counting sequence (Lesson 7 Template)

Note: This fluency activity reviews the grade-level standard of mentally finding 10 more or 10 less than a number without having to count.

Display the vertical counting sequence chart for reference.

- T: Say the number that is 1 more. 5. (Pause, and then snap.)
- S: 6.
- T: 15. (Pause, and then snap.)
- S: 16.

Continue with the following suggested sequence, as time permits: 55, 75, 105, 115; 67, 97, 107; 9, 49, 99, 109, 119.

> **NOTES ON MULTIPLE MEANS OF REPRESENTATION:**
>
> Some students may need practice writing the number after they mentally find 1 or 10 more or less. As an alternative for certain students, have them write the numbers instead of saying them.

Repeat for 10 more:
10, 40, 90, 100
3, 23, 63
56, 86, 96

Repeat for 10 less:
20, 50, 70
45, 65, 95
88, 118, 108

Repeat for 1 less:
4, 14, 84
8, 38, 88
10, 70, 120

Concept Development (31 minutes)

Materials: (T) 100-bead Rekenrek and 20-bead Rekenrek (if available), place value chart (Lesson 3 Template 2), personal white board, document camera (S) Place value chart (Lesson 3 Template 2), personal white board

Note: If the 20-bead Rekenrek is not available, draw two rows of large dots (5 white and 5 red in each row) on chart paper to represent two more rows of beads. Along with the bead sets, put the place value chart in a personal white board under the document camera, or put an image of the place value chart on an interactive board.

Gather students in the meeting area for today's lesson.

- T: You did a great job with counting the Say Ten way between the two Sprints today. Let's count by tens the Say Ten way. (Move the beads over on the Rekenrek as students count.)
- S: 1 ten, 2 tens, 3 tens, ..., 9 tens, 10 tens.
- T: (Write 10 in the tens position on the place value chart.) Since we were only counting tens, there are no additional ones, just 10 tens. (Write 0 in the ones position on the place value chart.)
- T: 10 tens is the same as ...?
- S: 100.

112   Lesson 8:   Count to 120 in unit form using only tens and ones. Represent numbers to 120 as tens and ones on the place value chart.

A STORY OF UNITS   Lesson 8 1•6

T: What if I add 1 more bead? (Hold up the 20-bead Rekenrek, and slide 1 bead over.) Do I still have 10 tens?
S: Yes!
T: But I also have…?
S: 1 one.
T: I need a volunteer to change our place value chart to show 10 tens and 1 one. (Select a student, and wait as she erases 0 in the ones position and writes 1.)
T: 10 tens 1 one is…?
S: 101. (Some students may say one hundred *and* one. If they do, explain that 100 + 1 describes 100 *and* 1, but the *name* of the number is one hundred one. This is similar to naming other numbers, such as 25. Twenty *and* 5 is written 20 + 5. To say the number, we say twenty-five.
T: We had 10 tens and then 10 tens 1. Next, we would have…? (Move another bead on the 20-bead Rekenrek.)
S: 10 tens 2.
T: Let's change our place value charts to record the tens and ones.
T: 10 tens 2 is the same as…?
S: 102.
T: Let's see. 100, 101, 102. Next would be…? (Slide a third bead.)
S: 103.
T: How many tens and ones are in 103? Let's change our place value charts to record the tens and ones.
T: Let's count together starting at 98. Listen for when I say to stop.
S/T: (Count together without the Rekenrek.) 98, 99, 100, 101, 102, 103, 104, 105, 106, 107, 108, 109.
T: Stop!
T: How many tens and ones are in 109? Talk with a partner. Let's show that many on the Rekenrek, and record it on your place value chart. (Circulate and notice students' recordings.)
T: Let's look at the Rekenrek. It shows how many tens?
S: 10 tens!
T: It shows how many additional ones?
S: 9 ones!
T: What if we slide over one more bead? How many tens would we have then?
S: 11 tens!
T: (Slide over one more bead so that the Rekenreks now show 11 tens.) Write this amount on your place value chart. Tell your partner what number has 11 tens. (Wait as students complete the task.)
T: 11 tens is the same as…?
S: One hundred ten!

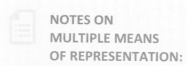

NOTES ON
MULTIPLE MEANS
OF REPRESENTATION:

If some students need more support, have them look at the Rekenreks as they count. This visual support will help them to identify the number of tens and then the number of additional ones.

Lesson 8:   Count to 120 in unit form using only tens and ones. Represent numbers to 120 as tens and ones on the place value chart.

A STORY OF UNITS
Lesson 8 1•6

Repeat the process, having students count from a given number and stop at a given number. Students identify the number in both its traditional form and its unit form. A suggested sequence would be 97 to 103, 108 to 112, and 108 to 120. Alternate between saying numbers the regular way and the Say Ten way. If students need more practice, the following partner activity may be used.

- Partner A uses quick tens and ones to draw a number between or including 100 and 120.
- Partner B records the number in the place value chart while Partner A writes the number below his drawing.
- The two partners check that they have matching numbers and then switch roles to start again.

Problem Set (10 minutes)

Students should do their personal best to complete the Problem Set within the allotted 10 minutes. For some classes, it may be appropriate to modify the assignment by specifying which problems they work on first. Some problems do not specify a method for solving. Students should solve these problems using the RDW approach used for Application Problems.

Student Debrief (10 minutes)

Lesson Objective: Count to 120 in unit form using only tens and ones. Represent numbers to 120 as tens and ones on the place value chart.

The Student Debrief is intended to invite reflection and active processing of the total lesson experience.

Invite students to review their solutions for the Problem Set. They should check work by comparing answers with a partner before going over answers as a class. Look for misconceptions or misunderstandings that can be addressed in the Debrief. Guide students in a conversation to debrief the Problem Set and process the lesson.

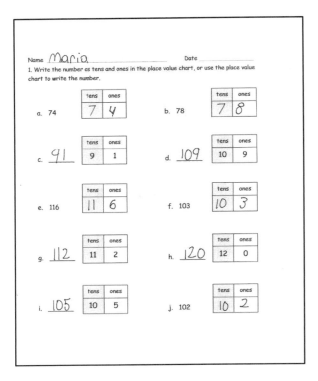

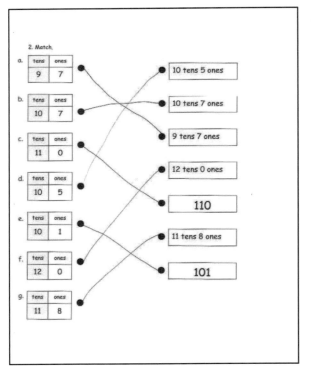

Lesson 8: Count to 120 in unit form using only tens and ones. Represent numbers to 120 as tens and ones on the place value chart.

EUREKA MATH

Any combination of the questions below may be used to lead the discussion.

- Look at Problem 1(d). What similarities and differences do you notice between reading a number and seeing the number in tens and ones?
- Look at Problem 2. Which matches were easy to identify, and which were more challenging? Explain why this was so.
- Choose a number from Problem 1. What is another way you could show this number in unit form? (This question is best used if students have been highly successful with today's lesson.)
- How can counting the Say Ten way help you with numbers from 100 to 120?
- Look at your Application Problem. Share your strategies for solving the problem.

Exit Ticket (3 minutes)

After the Student Debrief, instruct students to complete the Exit Ticket. A review of their work will help with assessing students' understanding of the concepts that were presented in today's lesson and planning more effectively for future lessons. The questions may be read aloud to the students.

Name _____   Date _____

1. Write the number as tens and ones in the place value chart, or use the place value chart to write the number.

a. 74

tens	ones

b. 78

tens	ones

c. _____

tens	ones
9	1

d. _____

tens	ones
10	9

e. 116

tens	ones

f. 103

tens	ones

g. _____

tens	ones
11	2

h. _____

tens	ones
12	0

i. _____

tens	ones
10	5

j. 102

tens	ones

Lesson 8: Count to 120 in unit form using only tens and ones. Represent numbers to 120 as tens and ones on the place value chart.

2. Match.

a. | tens | ones |
    |------|------|
    | 9 | 7 |

b. | tens | ones |
    |------|------|
    | 10 | 7 |

c. | tens | ones |
    |------|------|
    | 11 | 0 |

d. | tens | ones |
    |------|------|
    | 10 | 5 |

e. | tens | ones |
    |------|------|
    | 10 | 1 |

f. | tens | ones |
    |------|------|
    | 12 | 0 |

g. | tens | ones |
    |------|------|
    | 11 | 8 |

10 tens 5 ones

10 tens 7 ones

9 tens 7 ones

12 tens 0 ones

110

11 tens 8 ones

101

Lesson 8: Count to 120 in unit form using only tens and ones. Represent numbers to 120 as tens and ones on the place value chart.

Name _____ Date _____

1. Write the number as tens and ones in the place value chart, or use the place value chart to write the number.

a. 83

tens	ones

b. _____

tens	ones
9	4

c. _____

tens	ones
11	5

d. 106

tens	ones

2. Write the number.

   a. 10 tens 2 ones is the number _____.

   b. 11 tens 4 ones is the number _____.

A STORY OF UNITS   Lesson 8 Homework 1•6

Name _____   Date _____

1. Write the number as tens and ones in the place value chart, or use the place value chart to write the number.

a. 81

tens	ones

b. 98

tens	ones

c. ____

tens	ones
11	7

d. ____

tens	ones
10	8

e. 104

tens	ones

f. 111

tens	ones

2. Write the number.

a. 9 tens 2 ones is the number _____.	b. 8 tens 4 ones is the number _____.
c. 11 tens 3 ones is the number _____.	d. 10 tens 9 ones is the number _____.
e. 10 tens 1 ones is the number _____.	f. 11 tens 6 ones is the number _____.

Lesson 8: Count to 120 in unit form using only tens and ones. Represent numbers to 120 as tens and ones on the place value chart.

119

3. Match.

a. | tens | ones |
   |------|------|
   | 10 | 2 |

b. | tens | ones |
   |------|------|
   | 9 | 5 |

c. | tens | ones |
   |------|------|
   | 11 | 4 |

d. | tens | ones |
   |------|------|
   | 11 | 0 |

e. | tens | ones |
   |------|------|
   | 10 | 8 |

f. | tens | ones |
   |------|------|
   | 10 | 0 |

g. | tens | ones |
   |------|------|
   | 11 | 8 |

- 11 tens 4 ones
- 9 tens 5 ones
- 11 tens 8 ones
- 11 tens 0 ones
- 102
- 10 tens 0 ones
- 108

Lesson 9

Objective: Represent up to 120 objects with a written numeral.

Suggested Lesson Structure

■ Application Problem   (5 minutes)
■ Fluency Practice   (14 minutes)
□ Concept Development   (31 minutes)
■ Student Debrief   (10 minutes)
   Total Time   **(60 minutes)**

Application Problem (5 minutes)

Emi and Julio together have 17 pet mice. How many mice might each child have?

Extension: Who has more, and how many more does that child have?

Note: Today's Application Problem practices decomposing a two-digit number and can have more than one correct answer. This work supports students' compositions and decompositions when they begin Topic C in Lesson 10. Students compose, decompose, and recompose various two-digit addends.

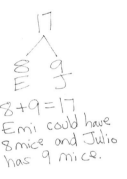

Fluency Practice (14 minutes)

- Sprint: +1, −1, +10, −10   **1.NBT.5**   (10 minutes)
- Beep-Counting   **1.NBT.1, 1.NBT.5**   (4 minutes)

Sprint: +1, −1, +10, −10 (10 minutes)

Materials: (S) +1, −1, +10, −10 Sprint

Note: This Sprint reviews the grade-level standard of mentally adding or subtracting 10 and supports students' understanding of place value.

Beep-Counting (4 minutes)

Note: This activity reviews counting and reading numbers to 120.

Write number sequences on the board with missing numbers. Students read the sequence aloud, saying "beep" for the missing number. Then, students say the missing number on the teacher's signal.

Lesson 9:   Represent up to 120 objects with a written numeral.   121

A STORY OF UNITS   **Lesson 9 1•6**

Use the following suggested sequence, as time permits:

a. 10, 11, 12, ___
b. 110, 111, 112, ___
c. 20, 19, 18, ___
d. 120, 119, 118, ___

e. 17, 18, ___, 20
f. 117, 118, ___, 120
g. 8, 9, ___, 11
h. 108, 109, ___, 111

i. 12, 11, ___, 9
j. 112, 111, ___, 109
k. ___, 7, 8, 9
l. ___, 107, 108, 109

Concept Development (31 minutes)

Materials: (T) 12 ten-sticks of linking cubes (ideally 6 red and 6 white ten-sticks), 10 additional loose linking cubes (S) Personal white board

Gather students with their personal white boards into a semicircle in the meeting area. The linking cubes should be placed close to the teacher but not in front of students.

T: Let's use our efficient counting skills to count different combinations of linking cubes. When I put out the linking cubes, your job is to count them as quickly as you can and write down the number of cubes I have. I put most of the cubes into sticks of ten, which should make it faster for you.

T: (Place 5 red ten-sticks and 5 white ten-sticks in the center for students to see. Scatter them far enough apart for students to count the 10 sticks. Wait as students count the sticks and record.)

T: How many linking cubes are here?
S: 100.

T: (Take all the sticks back. Place 10 ten-sticks down again, this time in a 5-group formation, with two rows of 5 sticks. Wait as students count and record. Check that students are recording 100 using the proper digits.)

T: How many linking cubes are here?
S: 100.

T: How did you know so quickly this time?
S: It's set up like 5 groups. → 5 tens and 5 tens is 10 tens. 10 tens is 100. → I saw 10 sets of sticks when I looked at them, so I knew 10 tens was 100.

T: (Lay out 12 ten-sticks using the 5-group formation with 2 more sticks on the side. As students count and record, watch for proper notation for 120.)

MP.4

T: How many tens do you see?
S: 12 tens.
T: How many cubes do you see?
S: 120 cubes.
T: How many ones would that be?
S: 120.

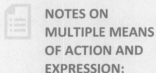

NOTES ON MULTIPLE MEANS OF ACTION AND EXPRESSION:

For students who are struggling, work together to write the number in a place value chart, and then check the placement of the digits in the number.

Lesson 9:   Represent up to 120 objects with a written numeral.

A STORY OF UNITS

Lesson 9 1•6

Repeat the process with the following number of linking cubes.

- 99
- 101
- 109
- 110
- 111
- 113
- 119
- 115
- 104
- 107
- 110 made with 10 ten-sticks and 10 additional separated ones

For more combinations, lay out objects for numbers between 98 and 120 using more than 10 ones, along with ten-sticks.

> **NOTES ON MULTIPLE MEANS OF ENGAGEMENT:**
>
> As a challenge for some students, use other combinations of tens and ones, such as 9 tens 16 ones.

Problem Set (10 minutes)

Students should do their personal best to complete the Problem Set within the allotted 10 minutes. For some classes, it may be appropriate to modify the assignment by specifying which problems they work on first. Some problems do not specify a method for solving. Students should solve these problems using the RDW approach used for Application Problems.

Student Debrief (10 minutes)

Lesson Objective: Represent up to 120 objects with a written numeral.

The Student Debrief is intended to invite reflection and active processing of the total lesson experience.

Invite students to review their solutions for the Problem Set. They should check work by comparing answers with a partner before going over answers as a class. Look for misconceptions or misunderstandings that can be addressed in the Debrief. Guide students in a conversation to debrief the Problem Set and process the lesson.

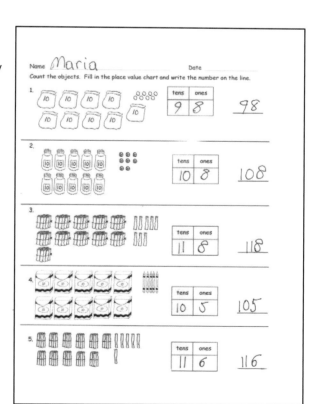

Lesson 9:   Represent up to 120 objects with a written numeral.

123

Any combination of the questions below may be used to lead the discussion.

- How many objects are in Problem 4? Problem 5? Which number is greater? Which picture takes up more space? What is another example of more objects taking up less space? Talk to your partner.
- Look at Problems 8 and 9. Which problem was quicker to draw and solve? Why?
- How is counting large numbers of objects like counting smaller numbers of objects? Explain your thinking. How is it different?
- Which beep-counting sequences are the quickest for you to answer? Why?
- Look at your Application Problem. What combinations did you use to show 17 pet mice? Are there other combinations that could be used?

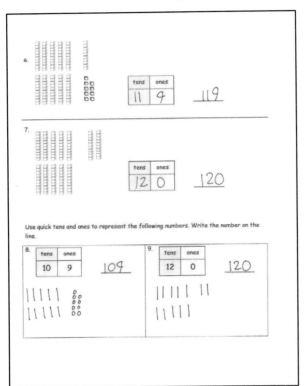

Exit Ticket (3 minutes)

After the Student Debrief, instruct students to complete the Exit Ticket. A review of their work will help with assessing students' understanding of the concepts that were presented in today's lesson and planning more effectively for future lessons. The questions may be read aloud to the students.

A

Lesson 9 Sprint 1•6

Name _____ Date _____

*Write the missing number. Pay attention to the addition or subtraction sign.

1.	5 + 1 = ☐		16.	29 + 10 = ☐
2.	15 + 1 = ☐		17.	9 + 1 = ☐
3.	25 + 1 = ☐		18.	19 + 1 = ☐
4.	5 + 10 = ☐		19.	29 + 1 = ☐
5.	15 + 10 = ☐		20.	39 + 1 = ☐
6.	25 + 10 = ☐		21.	40 − 1 = ☐
7.	8 − 1 = ☐		22.	30 − 1 = ☐
8.	18 − 1 = ☐		23.	20 − 1 = ☐
9.	28 − 1 = ☐		24.	20 + ☐ = 21
10.	38 − 1 = ☐		25.	20 + ☐ = 30
11.	38 − 10 = ☐		26.	27 + ☐ = 37
12.	28 − 10 = ☐		27.	27 + ☐ = 28
13.	18 − 10 = ☐		28.	☐ + 10 = 34
14.	9 + 10 = ☐		29.	☐ − 10 = 14
15.	19 + 10 = ☐		30.	☐ − 10 = 24

Lesson 9: Represent up to 120 objects with a written numeral.

B

Name _____ Date _____

*Write the missing number. Pay attention to the addition or subtraction sign.

1.	4 + 1 = ☐		16.	28 + 10 = ☐	
2.	14 + 1 = ☐		17.	9 + 1 = ☐	
3.	24 + 1 = ☐		18.	19 + 1 = ☐	
4.	6 + 10 = ☐		19.	29 + 1 = ☐	
5.	16 + 10 = ☐		20.	39 + 1 = ☐	
6.	26 + 10 = ☐		21.	40 − 1 = ☐	
7.	7 − 1 = ☐		22.	30 − 1 = ☐	
8.	17 − 1 = ☐		23.	20 − 1 = ☐	
9.	27 − 1 = ☐		24.	10 + ☐ = 11	
10.	37 − 1 = ☐		25.	10 + ☐ = 20	
11.	37 − 10 = ☐		26.	22 + ☐ = 32	
12.	27 − 10 = ☐		27.	22 + ☐ = 23	
13.	17 − 10 = ☐		28.	☐ + 10 = 39	
14.	8 + 10 = ☐		29.	☐ − 10 = 19	
15.	18 + 10 = ☐		30.	☐ − 10 = 29	

A STORY OF UNITS

Lesson 9 Problem Set 1•6

Name _____ Date _____

Count the objects. Fill in the place value chart, and write the number on the line.

1.

tens	ones

_____

2.

tens	ones

_____

3.

tens	ones

_____

4.

tens	ones

_____

5.

tens	ones

_____

Lesson 9: Represent up to 120 objects with a written numeral.

127

6.

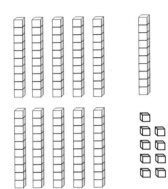

tens	ones

_____

7.

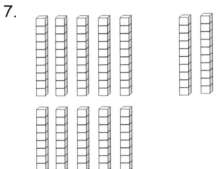

tens	ones

_____

Use quick tens and ones to represent the following numbers. Write the number on the line.

8. _____
tens	ones
10	9

9. _____
tens	ones
12	0

Name _____ Date _____

1. Count the objects. Fill in the place value chart, and write the number on the line.

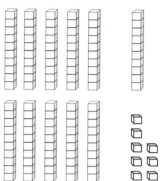

tens	ones

_____

2. Use quick tens and ones to represent the following numbers. Write the number on the line.

a.
tens	ones
11	0

_____

b.
tens	ones
10	1

_____

A STORY OF UNITS

Lesson 9 Homework 1•6

Name _____ Date _____

Count the objects. Fill in the place value chart, and write the number on the line.

1. [9 bags of 10 and 6 peanuts]

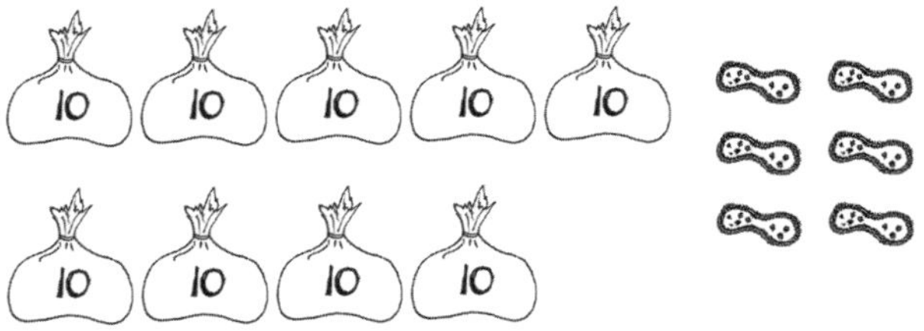

tens	ones

_____

2. [10 boxes of 10 crayons and 6 crayons]

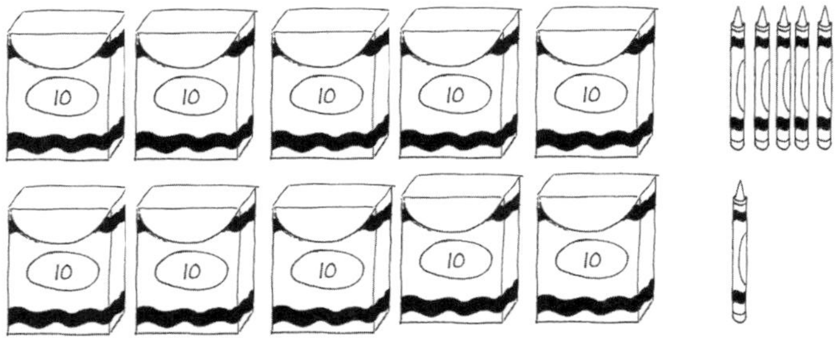

tens	ones

_____

3. [11 bags of 10 and 6 circles]

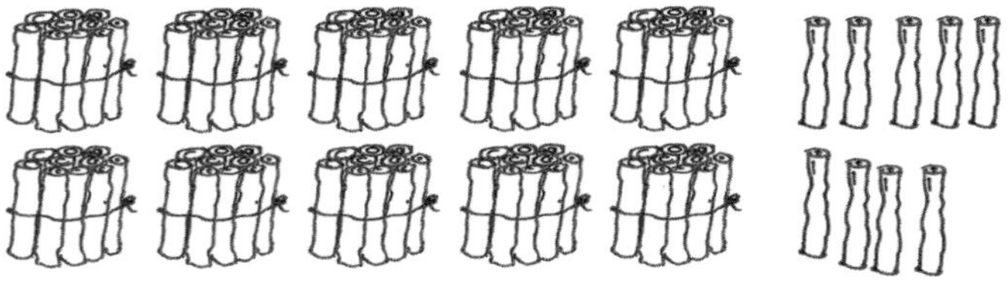

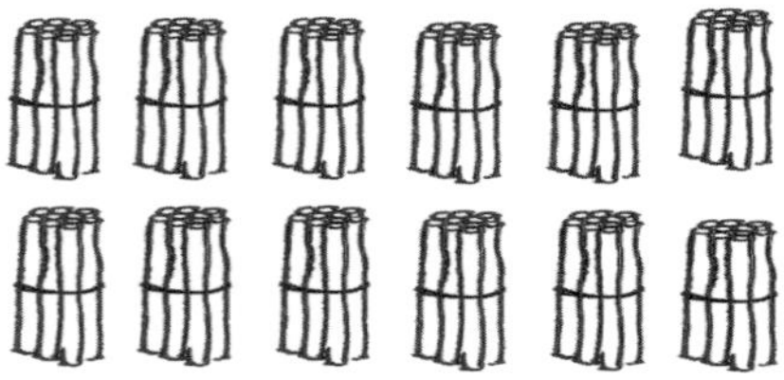

tens	ones

_____

4. [10 bundles of 10 and 7 sticks]

tens	ones

_____

5. [12 bundles of 10]

tens	ones

_____

Lesson 9: Represent up to 120 objects with a written numeral.

6.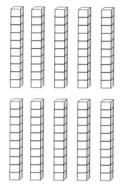

tens	ones

_____

7.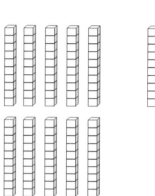

tens	ones

_____

Use quick tens and ones to represent the following numbers.
Write the number on the line.

8. _____

tens	ones
11	0

9. _____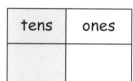

tens	ones
10	5

A STORY OF UNITS

Mathematics Curriculum

GRADE 1 • MODULE 6

Topic C
Addition to 100 Using Place Value Understanding

1.NBT.4, 1.NBT.6

Focus Standards:	1.NBT.4	Add within 100, including adding a two-digit number and a one-digit number, and adding a two-digit number and a multiple of 10, using concrete models or drawings and strategies based on place value, properties of operations, and/or the relationship between addition and subtraction; relate the strategy to a written method and explain the reasoning used. Understand that in adding two-digit numbers, one adds tens and tens, ones and ones; and sometimes it is necessary to compose a ten.
	1.NBT.6	Subtract multiples of 10 in the range 10–90 from multiples of 10 in the range 10–90 (positive or zero differences), using concrete models or drawings and strategies based on place value, properties of operations, and/or the relationship between addition and subtraction; relate the strategy to a written method and explain the reasoning used.
Instructional Days:	8	
Coherence -Links from:	G1–M4	Place Value, Comparison, Addition and Subtraction to 40
-Links to:	G2–M3	Place Value, Counting, and Comparison of Numbers to 1,000

During Topic C, students apply all of their place value and Level 3 strategy knowledge to add pairs of two-digit numbers to sums within 100. To this point, students have only added pairs of two-digit numbers within 40. They now extend their skills and strategies to larger pairs, such as 36 + 57, using all of the same methods.

Lesson 10 focuses students on number work with tens as they add and subtract multiples of 10 from multiples of 10. Students see that 20 + 70 is the same as 2 tens + 7 tens and that 80 – 50 is the same as 8 tens – 5 tens (**1.NBT.4**, **1.NBT.6**).

Building from student work with multiples of 10, Lesson 11 scaffolds students to add a multiple of 10 to any two-digit number, such as 64 + 30 (**1.NBT.4**). While some students may initially apply their ability to mentally add 10 by counting on by tens (64, 74, 84, 94), students also decompose 64 into 60 and 4 to solve, as shown to the right.

$$64 + 30 = 94$$
$$\diagup\diagdown$$
$$460$$
$$60 + 30 = 90$$
$$90 + 4 = 94$$

A STORY OF UNITS   Topic C 1•6

In Lesson 12, students add a pair of two-digit numbers when the ones digits have a sum less than or equal to 10 (**1.NBT.4**). They continue using strategies developed in Module 4. For example, when adding 47 + 23, students may choose to decompose the second addend into 20 and 3. They then add 20 to 47, making 67, and then add the remaining ones. Other students may choose to add the ones to the first addend and then add on the remaining tens, as shown to the right.

Lessons 13 and 14 focus on the most challenging addition work of this grade level as students add a pair of two-digit numbers when the ones digits have a sum greater than 10, as shown to the right (**1.NBT.4**).

During Lesson 15, students see how they can align materials or drawings to more distinctly separate and add tens with tens and ones with ones, recording the total below the drawings. Students connect this work with their decomposition work from Lessons 10 and 11, as shown to the right.

Lesson 16 extends this work, having students add a pair of two-digit numbers, such as 36 + 57, recording the 13 as 1 ten 3 ones as a part of their written method for recording their process. During Lesson 17, students continue to strengthen their skills and strategies to solve double-digit addition problems (**1.NBT.4**).

A Teaching Sequence Toward Mastery of Addition to 100 Using Place Value Understanding
Objective 1: Add and subtract multiples of 10 from multiples of 10 to 100, including dimes. (Lesson 10)
Objective 2: Add a multiple of 10 to any two-digit number within 100. (Lesson 11)
Objective 3: Add a pair of two-digit numbers when the ones digits have a sum less than or equal to 10. (Lesson 12)
Objective 4: Add a pair of two-digit numbers when the ones digits have a sum greater than 10 using decomposition. (Lessons 13–14)
Objective 5: Add a pair of two-digit numbers when the ones digits have a sum greater than 10 with drawing. Record the total below. (Lesson 15)
Objective 6: Add a pair of two-digit numbers when the ones digits have a sum greater than 10 with drawing. Record the new ten below. (Lessons 16–17)

Topic C: Addition to 100 Using Place Value Understanding   133

| A STORY OF UNITS | Lesson 10 1•6 |

Lesson 10

Objective: Add and subtract multiples of 10 from multiples of 10 to 100, including dimes.

Suggested Lesson Structure

■ Application Problem   (5 minutes)
■ Fluency Practice   (13 minutes)
■ Concept Development   (32 minutes)
■ Student Debrief   (10 minutes)
   Total Time   **(60 minutes)**

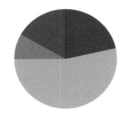

Application Problem (5 minutes)

Fran had 8 lizards. Anton gave some lizards to Fran. Fran now has 13 lizards. How many lizards did Anton give Fran?

Note: Today's problem is an *add to with change unknown* problem type. Some students may use a double tape diagram to solve, while others may choose to use a single tape diagram to solve.

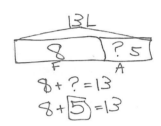

Fluency Practice (13 minutes)

- Core Fluency Differentiated Practice Sets **1.OA.6**   (5 minutes)
- Race to the Top! **1.OA.6**   (5 minutes)
- Get to Ten(s) **1.NBT.4**   (3 minutes)

Core Fluency Differentiated Practice Sets (5 minutes)

Materials: (S) Core Fluency Practice Sets (Lesson 1)

Note: Give the appropriate Practice Set to each student. Students who completed all questions correctly on their most recent Practice Set should be given the next level of difficulty. All other students should try to improve their scores on their current levels.

Students complete as many problems as they can in 90 seconds. Assign a counting pattern and start number for early finishers, or have them practice make ten addition or subtraction on the back of their papers. Collect and correct any Practice Sets completed within the allotted time.

A STORY OF UNITS   Lesson 10 1•6

Race to the Top! (5 minutes)

Materials: (S) Personal white board, Race to the Top! (Fluency Template), 2 dice per pair of students

Note: This fluency activity primarily targets the core fluency for Grade 1. Remember to closely monitor the strategies of students who are not performing well on the Practice Sets or Sprints. For students whose fine motor skills are not well developed, activities like Race to the Top! allow them to demonstrate their growing fluency.

Assign partners. Students take turns rolling the dice, saying an addition sentence, and recording the sums on the graph. The game ends when time runs out or one of the columns reaches the top of the graph.

Get to Ten(s) (3 minutes)

Materials: (T) 100-bead Rekenrek

Note: In this fluency activity, students apply their knowledge of partners to ten to find analogous partners to multiples of 10. Students need this skill when they learn to apply the make ten strategy to add two two-digit numbers in Lesson 13.

Model with the Rekenrek for the first few problems. Then, put the Rekenrek away to give students practice mentally getting to the next ten.

- T: (Show 9.) Say the number.
- S: 9.
- T: Say the number sentence to make ten.
- S: 9 + 1 = 10.
- T: (Move 1 bead to make 10. Show 19.)
- T: Say the number.
- S: 19.
- T: Say the number sentence to make 20.
- S: 19 + 1 = 20.

Continue with the following suggested sequence: 59, 79, 99; 5, 65, 85, 95; 8, 48, 78, 98; and 7, 37, 87, 97.

Concept Development (32 minutes)

Materials: (T) Chart paper, 10 dimes (S) Personal white board, number bond/number sentence set (Template), 5 dimes

Students sit in the meeting area in a semicircle formation.

- T: (Write 4 + 3 on the chart. Call up two volunteers.) Using your magic counting sticks, show us 4 + 3.
- S: (Student A shows 4 fingers; Student B shows 3 fingers.)

Lesson 10: Add and subtract multiples of 10 from multiples of 10 to 100, including dimes.

A STORY OF UNITS Lesson 10 1•6

T: How many fingers are there? Say the number sentence.
S: 4 + 3 = 7.
T: (Complete the number sentence on the chart.) Yes. 4 fingers + 3 fingers = 7 fingers.

On their personal white boards, have students write the number sentence, use math drawings to show 4 + 3 = 7, and make a number bond while recording the information on a chart.

T: Let's pretend these circles stand for bananas! Say the number sentence using bananas as the unit.
S: 4 bananas + 3 bananas = 7 bananas.
T: (Call for five additional volunteers to join the two volunteers.) Show us 4 tens + 3 tens using your magic counting sticks.
S: (Clasp hands to show 4 tens and 3 tens.)
T: (Help the first four students stand closer together to show 4 tens.)
T: (Point to the first four students.) How many tens do we have here?
S: 4 tens.
T: (Point to the last three students closely standing next to each other.) How many tens do we have here?
S: 3 tens.
T: How many tens are there in all?
S: 7 tens.
T: Say the number sentence the Say Ten way. (If students struggle, say, "Say the number sentence starting with 4 *tens*.")
S: 4 tens + 3 tens = 7 tens.
T: Say the number sentence the regular way starting with 40.
S: 40 + 30 = 70.
T: (Record the number sentence on the chart.)
T: (Point to the first problem on the chart.) Hmmm, how can knowing 4 + 3 = 7 help us with 4 tens + 3 tens? Turn and talk to your partner.
S: 4 tens + 3 tens = 7 tens is just like 4 + 3 = 7. It's just 4 things and 3 things make 7 things → 4 fingers and 3 fingers make 7 fingers. 4 bananas and 3 bananas make 7 bananas. 4 tens and 3 tens make 7 tens.
T: The numbers stay the same. The numbers, 4 and 3 and 7, stay the same, but the *units* change.

Direct students to write the number sentence, use math drawings, and make a number bond while charting their responses as shown to the bottom right.

NOTES ON MULTIPLE MEANS OF REPRESENTATION:

Students demonstrate a true understanding of math concepts when they can apply them in a variety of situations. Some students may not be able to make the connection between different number bonds as seen in this lesson. Their path to abstract thinking may be a little longer than others. Support these students with the use of manipulatives (linking cubes and coins) and plenty of practice on their personal white boards.

4 + 3 = 7   7
oooo + ooo   / \
                4   3

4 tens + 3 tens = 7 tens
|||| + ||| 7 tens
                    / \
                 4 tens 3 tens

40 + 30 = 70   70
                   / \
                  40 30

136   Lesson 10:   Add and subtract multiples of 10 from multiples of 10 to 100, including dimes.

A STORY OF UNITS

Lesson 10   1•6

Repeat the process using the following suggested sequence, and have students solve each problem using the Say Ten way and the regular way:

- 7 tens – 4 tens
- 30 + 60
- 9 dimes – 3 dimes
- 60 cents + 20 cents
- 70 + 30
- 10 tens – 4 tens

T: (Write 6 dimes – 4 dimes on the chart.) Draw a number bond for this subtraction problem, and share your thinking with your partner.

S: 6 dimes is the total. 4 dimes is one of the parts. → We know one part. The mystery is the other part to make 6 dimes or 60 cents → 6 dimes take away 4 dimes is 2 dimes → 60 cents take away 40 cents is 20 cents → I can take away a part from the total to find the missing part. (Show the number bond with 2 dimes still missing.)

T: What addition sentence can we write to match this number bond? Remember, we can say "unknown" or "mystery number."

S: 4 dimes + the mystery number = 6 dimes. (Record on the chart.)

T: What is the missing part?

S: 2 dimes!

T: Say the subtraction sentence and the related addition sentence the Say Ten way.

S: 6 tens – 4 tens = 2 tens. 4 tens + 2 tens = 6 tens.

T: Let's say it the regular way, too.

S: 60 – 40 = 20. 40 + 20 = 60.

Repeat the process as needed to support students' understanding.

Problem Set (10 minutes)

Students should do their personal best to complete the Problem Set within the allotted 10 minutes. For some classes, it may be appropriate to modify the assignment by specifying which problems they work on first. Some problems do not specify a method for solving. Students should solve these problems using the RDW approach used for Application Problems.

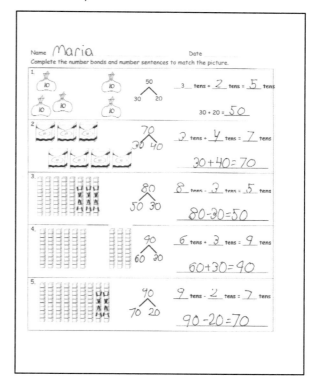

Lesson 10: Add and subtract multiples of 10 from multiples of 10 to 100, including dimes.

137

Lesson 10

Student Debrief (10 minutes)

Lesson Objective: Add and subtract multiples of 10 from multiples of 10 to 100, including dimes.

The Student Debrief is intended to invite reflection and active processing of the total lesson experience.

Invite students to review their solutions for the Problem Set. They should check work by comparing answers with a partner before going over answers as a class. Look for misconceptions or misunderstandings that can be addressed in the Debrief. Guide students in a conversation to debrief the Problem Set and process the lesson.

Any combination of the questions below may be used to lead the discussion.

- Look at Problems 1 and 2. Did you show your bonds the regular way or the Say Ten way?
- What did you notice about Problems 6 and 7? Can you find another set of problems that show a similar pattern?
- Using Problem 10, create a related problem by drawing a picture and writing the number sentence in the same way that Problems 6 and 7 go together.
- Write all the ways you can make a total of 10 tens or 100 using only tens. You may use three addends!
- Explain how knowing 3 + 6 can help solve 30 + 60.
- How can Race to the Top! and the Core Fluency Practice Sets help you solve addition and subtraction problems from today's lesson?

Exit Ticket (3 minutes)

After the Student Debrief, instruct students to complete the Exit Ticket. A review of their work will help with assessing students' understanding of the concepts that were presented in today's lesson and planning more effectively for future lessons. The questions may be read aloud to the students.

A STORY OF UNITS   Lesson 10 Problem Set 1•6

Name _____ Date _____

Complete the number bonds and number sentences to match the picture.

1.    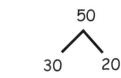   __3__ tens + ____ tens = ____ tens

   30 + 20 = _____

2. 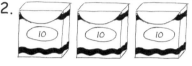   ____ tens + ____ tens = ____ tens

   _____

3. 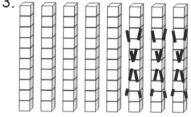   ⋀   ____ tens − ____ tens = ____ tens

   _____

4. 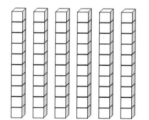 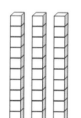   ⋀   ____ tens + ____ tens = ____ tens

   _____

5. 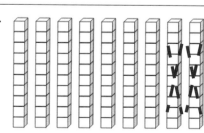   ____ tens − ____ tens = ____ tens

   _____

Lesson 10: Add and subtract multiples of 10 from multiples of 10 to 100, including dimes.

Count the dimes to add or subtract. Write a number sentence to match the value of the dimes.

6. +   40 + 20 = _____

7. ⊗ ⊗   _____

8. +   _____

9. ⊗   _____
    ⊗ ⊗

10.   _____

11. Fill in the missing numbers.

   a. 40 + 40 = _____   b. 50 – 30 = _____   c. 10 + _____ = 70

   d. 60 – _____ = 0   e. 90 – _____ = 10   f. 70 + _____ = 90

   g. 50 + 40 = _____   h. 100 – 30 = _____   i. 100 – _____ = 70

A STORY OF UNITS   Lesson 10 Exit Ticket 1•6

Name _____   Date _____

1. Fill in the missing numbers.

   a. 40 + 50 = _____   b. 80 – 60 = _____   c. 30 + _____ = 70

2. Write a number sentence to match the picture.

                    _____

Lesson 10:   Add and subtract multiples of 10 from multiples of 10 to 100, including dimes.

Name _____ Date _____

1. Complete the number bond or number sentence, and draw a line to the matching picture.

a.

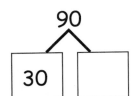

b.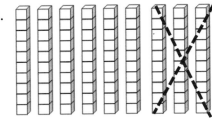

_____ − 40 = 60

c.

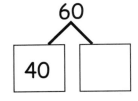

d.

80 − _____ = 60

2. Count the dimes to add or subtract. Write a number sentence to match the dimes.

a. +   40 + 20 = _____

b.    _____

c.    _____

d.    _____

3. Fill in the missing numbers.

a. 70 + _____ = 90   b. _____ + 30 = 80   c. 100 − _____ = 20

d. 30 + 60 = _____   e. 70 − _____ = 20   f. 20 + _____ = 60

g. _____ − 20 = 60   h. 90 − _____ = 20   i. 50 + _____ = 100

A STORY OF UNITS — Lesson 10 Fluency Template 1•6

Names _____ Date _____

 Race to the Top!

| 2 | 3 | 4 | 5 | 6 | 7 | 8 | 9 | 10 | 11 | 12 |

race to the top

144   Lesson 10:   Add and subtract multiples of 10 from multiples of 10 to 100, including dimes.

A STORY OF UNITS — Lesson 10 Template — 1•6

___ ◯ ___ ◯ ___

____ tens ◯ ____ tens ◯ ____ tens

___ ◯ ___ ◯ ___

number bond/number sentence set

Lesson 10: Add and subtract multiples of 10 from multiples of 10 to 100, including dimes.

Lesson 11

Objective: Add a multiple of 10 to any two-digit number within 100.

Suggested Lesson Structure

- ■ Application Problem   (5 minutes)
- ■ Fluency Practice   (10 minutes)
- ■ Concept Development   (35 minutes)
- ■ Student Debrief   (10 minutes)

  Total Time   **(60 minutes)**

Application Problem (5 minutes)

Ben sharpened 5 pencils. He has 8 more unsharpened pencils than sharpened pencils. How many unsharpened pencils does Ben have?

Note: Today's *comparison with bigger unknown* poses the additional challenge that there is only one person in the story. If students are still struggling with comparison problem types, consider altering the problem so that two students' pencils are being compared. Have a brief student discussion of the solution before moving on to the Fluency Practice.

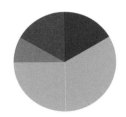

Fluency Practice (10 minutes)

- Core Fluency Differentiated Practice Sets **1.OA.6**   (5 minutes)
- Coin Drop **1.NBT.5, 1.MD.3**   (3 minutes)
- Get to the Next Ten **1.NBT.4**   (2 minutes)

Core Fluency Differentiated Practice Sets (5 minutes)

Materials: (S) Core Fluency Practice Sets (Lesson 1)

Note: Give the appropriate Practice Set to each student. Help students become aware of their improvement. After students do today's Practice Sets, ask them to stand if they tried a new level today or improved their score from the previous day. Consider having students clap once for each person standing to celebrate improvement.

A STORY OF UNITS
Lesson 11 1•6

Students complete as many problems as they can in 90 seconds. Assign a counting pattern and start number for early finishers, or have them practice make ten addition or subtraction on the back of their papers. Collect and correct any Practice Sets completed within the allotted time.

Coin Drop (3 minutes)

Materials: (T) 10 dimes, 10 pennies, can

Note: This activity reviews yesterday's lesson (Lesson 10), where students added and subtracted tens within 100.

Repeat the process from Lesson 5. Now that students have learned to add and subtract multiples of 10 from multiples of 10, the teacher may take out more than one dime at a time and have students calculate the remaining dimes.

Get to the Next Ten (2 minutes)

Note: This fluency activity builds on Lesson 10's Get to Ten(s) activity to prepare students for Lesson 13.

Say a number. Students say an addition sentence to get to the next multiple of 10. For the first few problems, begin with a number from 0 to 9 to provide students with a helper problem on which to build. Then, say numbers without providing the helper problem.

> T: Say the addition sentence to get to the next ten. 9.
> S: 9 + 1 = 10.
> T: 59.
> S: 59 + 1 = 60.

Continue with the following suggested sequence: 5, 65; 8, 78; 7, 87; and 6, 96.

Concept Development (35 minutes)

Materials: (T) 100-bead Rekenrek (S) Personal white board

Have students gather in the meeting area in a semicircle formation with their materials.

> T: (Write 40 + 30 = ? on chart paper.) On your personal white board, write the number sentence, and replace the question mark with the missing number. (Wait as students complete the task.)
> T: 40 + 30 is...?
> S: 70.

NOTES ON MULTIPLE MEANS OF ENGAGEMENT:

At this point in the year, students should be able to add a multiple of ten to any multiple of 10 within 100. If some students are struggling, have them use linking cubes in ten-sticks or quick ten drawings for more concrete or pictorial supports. Use the language of place value so that the dialogue begins to become part of their independent thinking. Work toward solving without the concrete supports.

Lesson 11: Add a multiple of 10 to any two-digit number within 100.   147

T: Explain how you know that 40 + 30 equals 70. You can draw or write on the chart paper to explain your thinking.

S: If you use the Rekenrek, you slide 4 tens over and then 3 tens over, and that's 7 tens, or 70. → Four tens plus 3 tens is 7 tens. That's 70. → In the place value chart, you add 3 tens to the 4 tens you have. (Post or show yesterday's chart paper, if available. Draw the place value chart and the number bond on today's chart paper.)

T: (Draw a line to start a new section of the chart paper. Write 45 + 30 = ? Move over 45 beads on the 100-bead Rekenrek.) On your personal white board, write this number sentence, and replace the question mark with the solution.

T: (Wait as students complete the task. If students do not know the answer right away, provide more time for them to remember solution strategies, e.g., quick ten drawings, the Rekenrek, counting on, decomposing, and composing).

T: 45 + 30 is…?

S: 75.

T: Who would like to share how they solved 45 + 30? Listen to your friends' ideas, and be ready to ask questions or comment. (As students are explaining, record their examples on the chart using number bonds and place value charts.)

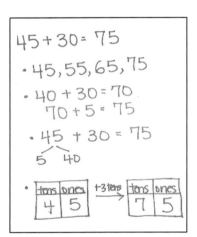

S: On the Rekenrek, there are 4 rows and 3 rows and 5 extra beads, so that's 7 tens and 5 ones, which is 75.

T: Does anyone have a question or comment about the Rekenrek solution?

S: Why did you say row? The five extras are a row, too.

S: Because I meant a row of ten. I guess I should say a full row.

T: Did anyone solve 45 + 30 in a different way?

S: I started at 45 and counted on ten 3 times. 45, 55, 65, 75.

T: Does anyone have a question or comment about the counting on solution?

S: Could you start counting on from 30?

S: Sure, I guess so. 30, 40, 50, 60, 70, 75. It's just easier for me the other way.

T: Did anyone solve 45 + 30 in a different way?

S: I broke 45 into 40 and 5 with the number bond, and then I added 40 and 30 first, which is 70, and then added on 5 to make 75.

T: Are there questions or comments about the number bond solution?

S: That's easy for me. I like that better than my way.

T: Why?

S: Because it's like I could just see it better. I counted on, and it seemed slower, too.

MP.3

NOTES ON MULTIPLE MEANS OF ACTION AND EXPRESSION:

Some students may get confused with all of the strategies available to them for solving problems. As the teacher, it might help these students to include one consistent method for solving. Then, students can share alternative strategies to allow exposure, but consistency really helps students who are struggling.

A STORY OF UNITS

Lesson 11 1•6

T: Did anyone solve 45 + 30 in a different way?
S: I thought of the place value chart and just added 3 tens to 4 tens and left the 5 ones alone. That gave me 75.
T: Are there comments and questions about the place value chart solution?
S: I don't understand what you mean that you left the 5 ones alone.
S: I mean when I was adding the tens, the ones didn't change.
T: It is important to really listen to your friends' solution strategies so that you can comment and ask questions.

Provide time for students to solve the following suggested sequence of problems. Students who would benefit from more concrete or pictorial support may use linking cubes in ten-sticks and ones, dimes and pennies, or quick ten drawings.

- 51 + 40
- 24 + 60
- 50 + 38
- 62 cents + 3 dimes
- 8 dimes + 12 cents
- 63 + ____ = 93
- 14 + ____ = 74
- ____ + 39 = 59
- ____ + 40 = 98

After each problem, have one or two students share a different method for solving the problem.

Problem Set (10 minutes)

Students should do their personal best to complete the Problem Set within the allotted 10 minutes. For some classes, it may be appropriate to modify the assignment by specifying which problems they work on first. Some problems do not specify a method for solving. Students should solve these problems using the RDW approach used for Application Problems.

Student Debrief (10 minutes)

Lesson Objective: Add a multiple of 10 to any two-digit number within 100.

The Student Debrief is intended to invite reflection and active processing of the total lesson experience.

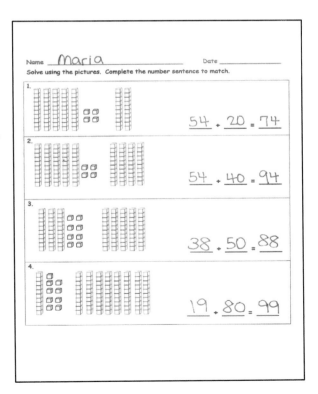

Lesson 11: Add a multiple of 10 to any two-digit number within 100.

149

A STORY OF UNITS

Lesson 11 1•6

Invite students to review their solutions for the Problem Set. They should check work by comparing answers with a partner before going over answers as a class. Look for misconceptions or misunderstandings that can be addressed in the Debrief. Guide students in a conversation to debrief the Problem Set and process the lesson.

Any combination of the questions below may be used to lead the discussion.

- Look at Problem 5 (c) and (d). How could solving Problem 5(c) help you solve Problem 5(d)?
- Look at Problem 6 (a) and (b). Did you or your partner use a different strategy than the number bond work from the top of the page? If so, explain your strategy.
- Look at Problem 6 (c) and (d). How did you find the missing addends? Explain your thinking.
- How is today's work similar to and different from yesterday's work?
- How did the coin drop fluency activity help you get better at adding tens?

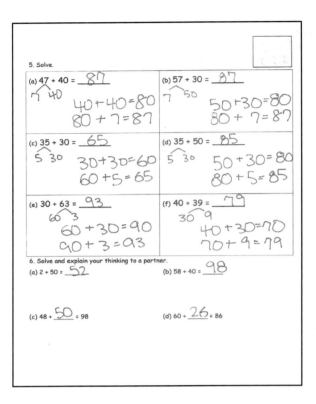

Exit Ticket (3 minutes)

After the Student Debrief, instruct students to complete the Exit Ticket. A review of their work will help with assessing students' understanding of the concepts that were presented in today's lesson and planning more effectively for future lessons. The questions may be read aloud to the students.

Lesson 11: Add a multiple of 10 to any two-digit number within 100.

Name _____ Date _____

Solve using the pictures. Complete the number sentence to match.

1.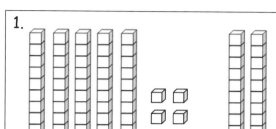

___ + ___ = ___

2.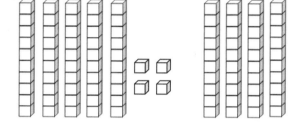

___ + ___ = ___

3.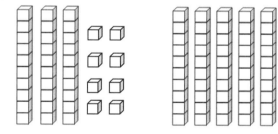

___ + ___ = ___

4.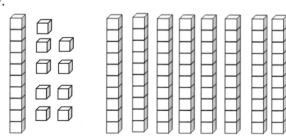

___ + ___ = ___

Lesson 11: Add a multiple of 10 to any two-digit number within 100.

A STORY OF UNITS   Lesson 11 Problem Set 1•6

5. Solve.

a. 47 + 40 = _____	b. 57 + 30 = _____
c. 35 + 30 = _____	d. 35 + 50 = _____
e. 30 + 63 = _____	f. 40 + 39 = _____

6. Solve and explain your thinking to a partner.

   a. 2 + 50 = _____   b. 58 + 40 = _____

   c. 48 + _____ = 98   d. 60 + _____ = 86

Name _____ Date _____

Solve. Use quick tens and ones drawings or number bonds.

| a. 42 + 50 = _____ | b. 30 + 57 = _____ |

Lesson 11: Add a multiple of 10 to any two-digit number within 100.

A STORY OF UNITS

Lesson 11 Homework 1•6

Name _____ Date _____

1. Solve using the pictures. Complete the number sentence to match.

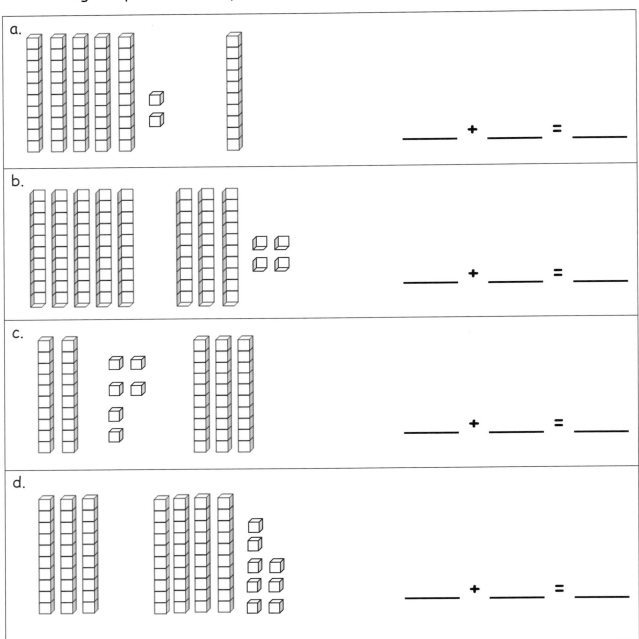

a. ____ + ____ = ____

b. ____ + ____ = ____

c. ____ + ____ = ____

d. ____ + ____ = ____

A STORY OF UNITS　　　　　　　　　　　　　　　　　　　　Lesson 11 Homework 1•6

$$64 + 30 = 94$$
$$\diagup\diagdown$$
$$4 \quad 60$$
$$60 + 30 = 90$$
$$90 + 4 = 94$$

2. Use number bonds to solve.

a. 38 + 40 = _____	b. 54 + 30 = _____
c. 46 + 40 = _____	d. 30 + 57 = _____
e. 20 + 68 = _____	f. 25 + 70 = _____

3. Solve. You may use number bonds to help you.

　　a. 72 + 20 = _____　　　　　　　　　　b. 48 + 50 = _____

　　c. 46 + _____ = 96　　　　　　　　　　d. _____ + 40 = 87

Lesson 11:　Add a multiple of 10 to any two-digit number within 100.

Lesson 12

Objective: Add a pair of two-digit numbers when the ones digits have a sum less than or equal to 10.

Suggested Lesson Structure

■ Application Problem   (5 minutes)
■ Fluency Practice   (15 minutes)
■ Concept Development   (30 minutes)
■ Student Debrief   (10 minutes)
  Total Time   **(60 minutes)**

Application Problem (5 minutes)

Kiana wants to have 14 stickers in her folder. She needs 6 more stickers to make her goal. How many stickers does she have right now?

Note: Today's problem is an *add to with start unknown* problem type. This can be challenging because some students associate the word *more* in a problem as meaning they must add.

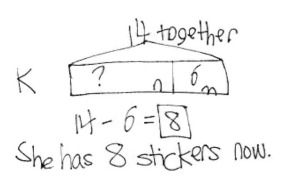

Fluency Practice (15 minutes)

- Grade 1 Core Fluency Sprint **1.OA.6**   (10 minutes)
- Add Tens **1.NBT.4**   (3 minutes)
- Analogous Addition Sentences **1.OA.6, 1.NBT.4** (2 minutes)

Grade 1 Core Fluency Sprint (10 minutes)

Materials: (S) Core Fluency Sprints (Lesson 3)

Note: Choose an appropriate Sprint based on the needs of the class. As students work, pay attention to their strategies and the number of problems they are answering. If the majority of students complete the first three quadrants today, try giving them the next level of difficulty when administering the next Sprint. If many students are not making it to the third quadrant, consider repeating today's Sprint.

Core Fluency Sprint List:

- Core Addition Sprint 1
- Core Addition Sprint 2
- Core Subtraction Sprint
- Core Fluency Sprint: Totals of 5, 6, and 7
- Core Fluency Sprint: Totals of 8, 9, and 10

Add Tens (3 minutes)

Materials: (S) Personal white board, die per pair of students

Note: This fluency activity reviews adding multiples of 10 to two-digit numbers, which helps prepare students for today's lesson.

Choose a student to help model the activity. Then, assign partners of equal ability to work together.

- Partner A writes or draws a number (with quick tens and ones) between 10 and 40 (e.g., 25).
- Partner B rolls the die to determine the number of tens to add (e.g., if she rolls 5, add 5 tens).
- Both partners write the number sentence on their personal white boards and check each other's work (e.g., 25 + 50 = 75).

Analogous Addition Sentences (2 minutes)

Note: This fluency activity encourages students to use sums within 10 to solve more challenging problems. Reviewing adding a one-digit number to a two-digit number when the ones have a sum less than or equal to 10 prepares students for today's lesson.

T: Say the number sentence with the answer. 3 + 2.
S: 3 + 2 = 5.
T: 43 + 2.
S: 43 + 2 = 45.
T: 42 + 3.
S: 42 + 3 = 45.
T: 3 + 42.
S: 3 + 42 = 45.

Continue with the following suggested sequence:

6 + 2	4 + 3	6 + 3
56 + 2	64 + 3	96 + 3
96 + 2	63 + 4	93 + 6
42 + 6	4 + 63	6 + 93

Lesson 12: Add a pair of two-digit numbers when the ones digits have a sum less than or equal to 10.

A STORY OF UNITS — Lesson 12 1•6

Concept Development (30 minutes)

Materials: (T) Chart paper (S) Personal white board

Begin today's lesson with students sitting at their desks or tables with their materials.

MP.5 Three sets of problems have been provided for students to extend their double-digit addition skills from Module 4. Choose the appropriate set, or portion of a set, that best meets students' needs. Although it may be tempting to begin with a review of a particular method to solve problems, refrain from doing so at the onset of the lesson. Instead, encourage and remind students of their toolkit: number sentences, the place value chart, linking cubes, drawings, number bonds, counting on, etc. Although students may ask questions, resist giving hints or solving the problem as a class. Continue, however, to ask questions that will prompt students to use their toolkit. For example, "How can this be solved? What method could you use?"

> **NOTES ON MULTIPLE MEANS OF ACTION AND EXPRESSION:**
>
> Some students may benefit from more concrete or pictorial supports. Use linking cubes in ten-sticks and ones as well as quick ten drawings for these students. While supporting students with these materials, be sure to connect them with number sentences with decomposed bonds to support increased understanding. See Module 4 Lesson 24 for examples of how these materials have been used for similar instructional objectives.

After each problem, have students share their solutions and invite one or two students to explain their strategies. Today, try to preselect students who have used varied strategies, such as adding ones first or adding tens first. Encourage students to use place value language to describe strategies for solving. Ask questions such as, "What is another way this can be solved? Why did you choose your method?"

$$23+57=80 \qquad 23+57=80$$
$$/\ \backslash \qquad\qquad\qquad \backslash\ /$$
$$3\ \ 20 \qquad\qquad\quad 20\ \ 3$$

$$57+20=77 \qquad 57+3=60$$
$$77+3=80 \qquad 60+20=80$$

In Problems 1–4, pairs of two-digit numbers from Module 4 Lessons 24 and 25 are presented with an analogous problem using numbers from 40 to 100 from Module 6.

Problems 5–8 provide a scaffold-less opportunity to add pairs of two-digit numbers.

Problems 9–12 encourage students to identify the missing number in varied positions within the number sentence.

Problems 1–4	Problems 5–8	Problems 9–12
24 + 13, then solve 54 + 13	76 + 23	63 + ____ = 84
15 + 13, then solve 45 + 23	23 + 57	48 + ____ = 100
15 + 15, then solve 45 + 45	41 + 39	____ + 59 = 70
26 + 14, then solve 66 + 34	34 + 53	32 + ____ = 100

158 Lesson 12: Add a pair of two-digit numbers when the ones digits have a sum less than or equal to 10.

A STORY OF UNITS   Lesson 12 1•6

Should students need additional support, the following dialogue presents a more guided approach to Problems 1–4.

T: (Write 24 + 10 on chart paper.) Use quick tens to show and solve this problem. (Wait as students draw on their personal white boards.)

T: 24 + 10 is…?

S: 34.

T: (Write 24 + 13 on chart paper.) Use quick tens to show and solve this problem. (Wait as students draw on their boards.)

T: 24 + 13 is…?

S: 37.

T: What did you do to solve this problem? Turn and talk with a partner. (Wait as students discuss.)

S: I took apart 13, making it 10 and 3. I added 10 first; that's 34, and then 3 more makes 37. → I already knew 24 + 10 was 34, so 3 more was 37.

T: (As students explain, use number bonds with number sentences to record their process.)

T: Great job adding the tens and then adding the rest of the ones.

T: (Write 54 + 13 on chart paper.) Solve this problem using your same thinking. If quick tens will help you, use them, or challenge yourself to use number bonds with your number sentence to solve the problem. (Wait as students draw on personal white boards.)

T: 54 + 13 is…?

S: 67.

NOTES ON MULTIPLE MEANS OF ACTION AND EXPRESSION:

Encourage students to explain their thinking about adding or subtracting tens. Students may learn as much from one another's reasoning as from the lesson. This also provides the opportunity for the teacher to learn more about a student's level of thinking and ability to express that thinking.

$$24 + 13 = 37$$
$$\overset{\wedge}{10\ 3}$$

$$54 + 13 = 67$$
$$\overset{\wedge}{10\ 3}$$

Invite students to share how they solved this problem. Emphasize their process of decomposing at least one number into tens and ones as they put the addends together. Repeat this process for 15 + 13 and 45 + 13. When beginning 15 + 15, note that students may choose to add the ones first as shown below.

T: (Write 15 + 15 on chart paper.) Solve this problem. (Wait as students solve.)

T: 15 + 15 is…?

S: 30.

T: What did you do to solve this problem?

S: I took apart the second 15, making it 10 and 5. I added 10 first; that's 25, and then 5 more makes it 30. → I started the same way, but I added 15 + 5 first; that's 20, and then I added 10 more to make 30. → I made both fifteens into 10 and 5. I added 5 and 5 to make 10, so then I had 3 tens. That's 30.

$$15 + 15 = 30 \qquad 15 + 15 = 30$$
$$\overset{\wedge}{10\ 5} \qquad \overset{\wedge}{5\ 10}$$
$$15 + 10 = 25 \qquad 15 + 5 = 20$$
$$25 + 5 = 30 \qquad 20 + 10 = 30$$

$$15 + 15 = 30$$
$$\overset{\wedge}{10\ 5}\ \overset{\wedge}{5\ 10}$$
$$5 + 5 = 10$$
$$10 + 10 + 10 = 30$$

Lesson 12: Add a pair of two-digit numbers when the ones digits have a sum less than or equal to 10.

A STORY OF UNITS

Lesson 12   1•6

Use number bonds and number sentences to record students' methods. If all students add the tens first, pose the other methods as ways that solved the problem, as an opportunity to consider alternative methods.

T: (Point to the example while describing each method.) Some of you broke the second 15 into tens and ones and added the tens first and then the ones. Some of you broke the second 15 into tens and ones and added the ones first and then the tens. A few of you broke both fifteens into tens and ones and added ones with ones and tens with tens. Did you all find the total of 30?

S: Yes!

Have students work on the following problems or repeat the same process with the following: 45 + 45, 26 + 14, and 66 + 34.

Problem Set (10 minutes)

Students should do their personal best to complete the Problem Set within the allotted 10 minutes. For some classes, it may be appropriate to modify the assignment by specifying which problems they work on first. Some problems do not specify a method for solving. Students should solve these problems using the RDW approach used for Application Problems.

Student Debrief (10 minutes)

Lesson Objective: Add a pair of two-digit numbers when the ones digits have a sum less than or equal to 10.

The Student Debrief is intended to invite reflection and active processing of the total lesson experience.

Invite students to review their solutions for the Problem Set. They should check work by comparing answers with a partner before going over answers as a class. Look for misconceptions or misunderstandings that can be addressed in the Debrief. Guide students in a conversation to debrief the Problem Set and process the lesson.

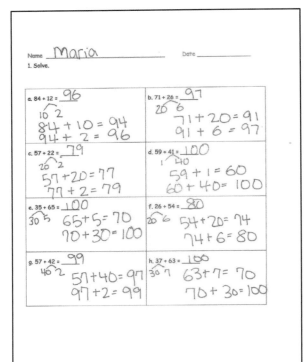

160   Lesson 12:   Add a pair of two-digit numbers when the ones digits have a sum less than or equal to 10.

EUREKA MATH

A STORY OF UNITS
Lesson 12 1•6

Any combination of the questions below may be used to lead the discussion.

- Look at Problem 1. Did you solve all of your problems the same way? What was your strategy?
- Did anyone solve some problems one way and then use a different strategy to solve other problems? Explain your reasoning.
- How does yesterday's work with adding multiples of 10 connect to today's work?
- How did your fluency work today help you with today's problems? Use specific examples to explain your thinking.
- Look at your Application Problem. Share your solution and your strategy for solving.

Exit Ticket (3 minutes)

After the Student Debrief, instruct students to complete the Exit Ticket. A review of their work will help with assessing students' understanding of the concepts that were presented in today's lesson and planning more effectively for future lessons. The questions may be read aloud to the students.

Lesson 12: Add a pair of two-digit numbers when the ones digits have a sum less than or equal to 10.

Name _____ Date _____

1. Solve.

a. 84 + 12 = _____	b. 71 + 26 = _____
c. 57 + 22 = _____	d. 59 + 41 = _____
e. 35 + 65 = _____	f. 26 + 54 = _____
g. 57 + 42 = _____	h. 37 + 63 = _____

A STORY OF UNITS

Lesson 12 Problem Set 1•6

2. Solve.

a. 45 + 13 = _____	b. 45 + 23 = _____
c. 21 + 27 = _____	d. 27 + 23 = _____
e. 48 + 32 = _____	f. 48 + 52 = _____
g. 34 + 65 = _____	h. 46 + 43 = _____

Lesson 12: Add a pair of two-digit numbers when the ones digits have a sum less than or equal to 10.

A STORY OF UNITS

Lesson 12 Exit Ticket 1•6

Name _____   Date _____

Solve using number bonds. You may choose to add the ones or tens first. Write the two number sentences to show what you did.

a. 56 + 43 = _____

b. 22 + 75 = _____

A STORY OF UNITS

Lesson 12 Homework 1•6

Name _____ Date _____

1. Solve.

a. 46 + 22 = _____	b. 74 + 23 = _____
c. 54 + 25 = _____	d. 68 + 31 = _____
e. 45 + 55 = _____	f. 86 + 13 = _____
g. 37 + 52 = _____	h. 47 + 52 = _____

Lesson 12: Add a pair of two-digit numbers when the ones digits have a sum less than or equal to 10.

A STORY OF UNITS

Lesson 12 Homework 1•6

2. Solve using number bonds. You may choose to add the ones or tens first. Write the two number sentences to show what you did.

a. 76 + 23 = _____	b. 45 + 33 = _____
c. 31 + 67 = _____	d. 57 + 32 = _____
e. 58 + 21 = _____	f. 25 + 63 = _____
g. 44 + 55 = _____	h. 47 + 53 = _____

Lesson 12: Add a pair of two-digit numbers when the ones digits have a sum less than or equal to 10.

Lesson 13

Objective: Add a pair of two-digit numbers when the ones digits have a sum greater than 10 using decomposition.

Suggested Lesson Structure

- ■ Application Problem (5 minutes)
- ■ Fluency Practice (14 minutes)
- ■ Concept Development (31 minutes)
- ■ Student Debrief (10 minutes)

Total Time **(60 minutes)**

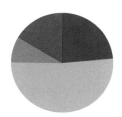

Application Problem (5 minutes)

Julio read 6 books this week. Emi read 12 books this week. How many fewer books did Julio read than Emi? How many books did they read in all? How many more books does Julio have to read so that he has read one more book than Emi?

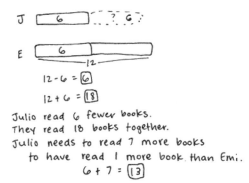

Note: Today's problem begins with a *comparison with difference unknown* problem type. Each of the succeeding questions can help students contrast this type of question with both the *put together with result unknown* problem type and the *add to with change unknown* problem type.

Fluency Practice (14 minutes)

- Grade 1 Core Fluency Sprint **1.OA.6**   (10 minutes)
- Make Ten Addition with Partners **1.OA.6**   (4 minutes)

Grade 1 Core Fluency Sprint (10 minutes)

Materials: (S) Core Fluency Sprints (Lesson 3)

Note: Based on the needs of the class, select a Sprint from Lesson 3's materials. There are several possible options available.

1. Re-administer the Sprint used during the previous lesson.
2. Administer the next Sprint in the sequence.
3. Differentiate. Administer two different Sprints. Simply have one group do a counting activity on the back of the Sprint while the other Sprint is corrected.

Lesson 13

Make Ten Addition with Partners (4 minutes)

Materials: (S) Personal white board

Note: This fluency activity reviews how to use the Level 3 strategy of making ten to add two single-digit numbers. Reviewing the make ten strategy prepares students for today's lesson, in which they make ten to add two two-digit numbers.

- Assign partners of equal ability.
- Partners choose an addend for each other from 1 to 10.
- On their personal white boards, students add their numbers to 9, 8, and 7. Remind students to write the two addition sentences they learned in Module 2.
- Partners then exchange boards and check each other's work.

NOTES ON MULTIPLE MEANS OF ENGAGEMENT:

Careful selection of pairs for collaborative work is essential to achieving expected outcomes. Some lessons lend themselves to groupings of students with similar skill sets, while others work better when students are heterogeneously grouped. Some students benefit from the opportunity to work independently and share with the teacher or another pair after they have completed the task.

Concept Development (31 minutes)

Materials: (T) Chart paper, document camera (if available)
(S) Personal white board

Gather students in the meeting area with their materials in a semicircle formation.

Three sets of problems extend students' double-digit addition skills from Module 4. Although it may be tempting to review a particular method to solve two-digit addition problems, refrain from doing so. Instead, encourage and remind students of the same tools they used in Lesson 12.

MP.5

After each problem, have students share their solutions. Invite one or two students to explain their strategies for solving. They may redraw their work or display the work using a document camera. Select work that represents a variety of strategies, including decomposing to get to the next ten, adding the tens and then the ones, and adding the ones and then the tens.

NOTES ON MULTIPLE MEANS OF ACTION AND EXPRESSION:

Students may choose how they want to solve problems—with drawings, number bonds, or the arrow way. Students should begin to move away from drawing to the more abstract method of problem solving. However, not all students are ready to abstractly solve problems, so support students wherever they are in their learning, and guide them as they progress.

Lesson 13: Add a pair of two-digit numbers when the ones digits have a sum greater than 10 using decomposition.

A STORY OF UNITS   Lesson 13   1•6

Encourage students to use place value language to describe how their strategy works. Ask questions such as, "Why did you choose your method?"

Problems 1–4 review work from Module 4 Lessons 26 and 27 with analogous problems now between 40 and 100.

Problems 5–12 provide a scaffold less opportunity to add pairs of two-digit numbers.

Problems 1–4

19 + 11, 59 + 11, 59 + 21

19 + 13, 59 + 13, 59 + 33

18 + 15, 68 + 25

17 + 16, 37 + 56

Problems 5–12

49 + 12

59 + 22

48 + 24

54 + 38

37 + 37

37 + 46

78 + 22

33 + 67

NOTES ON MULTIPLE MEANS OF ACTION AND EXPRESSION:

Continue to challenge students working above grade level. Change some of the expressions into number sentences with missing addends, or give students some word problems to solve with similar numbers.

Should students need additional support, the following dialogue presents a more guided approach to Problems 2–4. Problem 1 (a), (b), and (c) practice the work from yesterday's lesson to segue into today's objective.

T: (Write 19 + 13 = ___ on the chart.) Use quick tens to show these two numbers. Then, solve for the total. (Circulate as students work to assess students' ability to solve independently and identify common errors.)

T: 19 plus 13 equals…?

S: 32.

T: Talk with your partner about how you solved the problem. Try to show your thinking using number bonds with your number sentence. (Circulate as students explain their solution methods and create written notation of their methods.)

T: I heard many of you say you started with 19 and added 10. (Select a student who used this method to show the class. Walk through the steps of breaking apart 13 into 10 and 3. 19 + 10 is 29. Then, to add 29 + 3, the student may have broken 3 into 1 and 2, for a total of 32, as shown to the right. If the student's written notation is appropriate, have her share her written notation. If it is not, then model the number sentence and number bond work as the student describes her process.)

T: 19 is so close to 20. You're all very good at adding multiples of ten. How could I break 13 to make the next ten and then add the rest? How much more does 19 need to make 20?

S: 1 more!

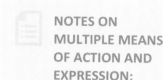

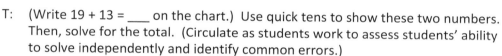

Lesson 13: Add a pair of two-digit numbers when the ones digits have a sum greater than 10 using decomposition.

169

A STORY OF UNITS **Lesson 13 1•6**

T: I would break 13 into 1 and...? (Begin written notation to show the bond below 13.)
S: 12.
T: Our first number sentence would be 19 + 1 is...?
S: 20.
T: Then, we would have...?
S: 20 + 12 = 32.

Repeat the process with the analogous problem of 59 + 13 and then with 59 + 33. When moving on to Problems 3 and 4, consider asking students to take on more of the demonstrations and explanations.

$$59+13=72 \qquad 59+13=72$$
$$\quad\;\;\wedge \qquad\qquad\qquad \wedge$$
$$\;\;10\;\;3 \qquad\qquad\;\; 1\;\;12$$
$$59+10=69 \qquad 59+1=60$$
$$69+3=72 \qquad 60+12=72$$
$$\quad\;\;\wedge$$
$$\;\;1\;\;2$$

Problem Set (10 minutes)

Students should do their personal best to complete the Problem Set within the allotted 10 minutes. For some classes, it may be appropriate to modify the assignment by specifying which problems they work on first. Some problems do not specify a method for solving. Students should solve these problems using the RDW approach used for Application Problems.

Student Debrief (10 minutes)

Lesson Objective: Add a pair of two-digit numbers when the ones digits have a sum greater than 10 using decomposition.

The Student Debrief is intended to invite reflection and active processing of the total lesson experience.

Invite students to review their solutions for the Problem Set. They should check work by comparing answers with a partner before going over answers as a class. Look for misconceptions or misunderstandings that can be addressed in the Debrief. Guide students in a conversation to debrief the Problem Set and process the lesson.

Any combination of the questions below may be used to lead the discussion.

- Which problem was the easiest for you to solve? What made it easy for you?
- Find two problems in your Problem Set that are related in some way. Explain your thinking.
- How is Make Ten Addition from today's Fluency Practice related to some of the work you did on your Problem Set?

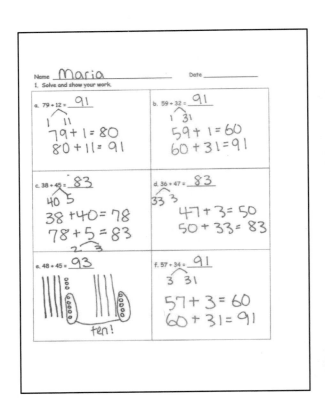

Lesson 13: Add a pair of two-digit numbers when the ones digits have a sum greater than 10 using decomposition.

Lesson 13

Exit Ticket (3 minutes)

After the Student Debrief, instruct students to complete the Exit Ticket. A review of their work will help with assessing students' understanding of the concepts that were presented in today's lesson and planning more effectively for future lessons. The questions may be read aloud to the students.

2. Solve and show your work.

a. 24 + 37 = 61

b. 48 + 45 = 93
40 8 40 5
40 + 40 = 80
8 + 5 = 13
80 + 13 = 93

c. 29 + 67 = 96
1 66
29 + 1 = 30
30 + 66 = 96

d. 48 + 34 = 82
30 4
48 + 30 = 78
78 + 4 = 82
2 2

e. 69 + 27 = 96
1 26
69 + 1 = 70
70 + 26 = 96

f. 78 + 17 = 95
70 8 10 7
70 + 10 = 80
8 + 7 = 15
80 + 15 = 95

Lesson 13: Add a pair of two-digit numbers when the ones digits have a sum greater than 10 using decomposition.

A STORY OF UNITS　　　　　　　　　　　　　　　　　　　　Lesson 13 Problem Set 1•6

Name _____   Date _____

1. Solve and show your work.

a. 79 + 12 = _____	b. 59 + 32 = _____
c. 38 + 45 = _____	d. 36 + 47 = _____
e. 48 + 45 = _____	f. 57 + 34 = _____

Lesson 13: Add a pair of two-digit numbers when the ones digits have a sum greater than 10 using decomposition.

2. Solve and show your work.

a. 24 + 37 = _____

b. 48 + 45 = _____

c. 29 + 67 = _____

d. 48 + 34 = _____

e. 69 + 27 = _____

f. 78 + 17 = _____

Lesson 13: Add a pair of two-digit numbers when the ones digits have a sum greater than 10 using decomposition.

Name _____ Date _____

Solve and show your work.

a. 49 + 37 = _____	b. 56 + 38 = _____

Lesson 13: Add a pair of two-digit numbers when the ones digits have a sum greater than 10 using decomposition.

A STORY OF UNITS

Lesson 13 Homework 1•6

Name _____ Date _____

1. Solve and show your work.

a. 15 + 26 = _____	b. 46 + 49 = _____	c. 28 + 54 = _____
d. 69 + 13 = _____	e. 69 + 23 = _____	f. 69 + 19 = _____
g. 49 + 43 = _____	h. 57 + 36 = _____	i. 68 + 23 = _____

Lesson 13: Add a pair of two-digit numbers when the ones digits have a sum greater than 10 using decomposition.

A STORY OF UNITS

Lesson 13 Homework 1•6

2. Solve and show your work.

a. 34 + 47 = _____	b. 38 + 45 = _____	c. 68 + 23 = _____
d. 39 + 57 = _____	e. 38 + 44 = _____	f. 17 + 76 = _____
g. 68 + 24 = _____	h. 18 + 77 = _____	i. 14 + 67 = _____

Lesson 13: Add a pair of two-digit numbers when the ones digits have a sum greater than 10 using decomposition.

EUREKA MATH

A STORY OF UNITS   Lesson 14 1•6

Lesson 14

Objective: Add a pair of two-digit numbers when the ones digits have a sum greater than 10 using decomposition.

Suggested Lesson Structure

■ Application Problem   (5 minutes)
■ Fluency Practice   (13 minutes)
□ Concept Development   (32 minutes)
■ Student Debrief   (10 minutes)
  Total Time   **(60 minutes)**

Application Problem (5 minutes)

There are 12 chairs at the lunch table and 15 students. How many more chairs are needed so that every student has a chair?

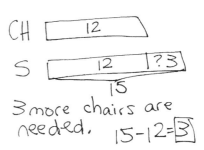

Note: Today's problem is a *comparison with difference unknown* problem type. Students who have struggled with comparison problems may successfully solve this common real-life problem. Before moving on to the Fluency Practice, have students briefly discuss the solution.

Fluency Practice (13 minutes)

- Core Fluency Differentiated Practice Sets **1.OA.6**   (5 minutes)
- Add Tens **1.NBT.4**   (3 minutes)
- Take Out Ones **1.OA.6, 1.NBT.4**   (5 minutes)

Core Fluency Differentiated Practice Sets (5 minutes)

Materials: (S) Core Fluency Practice Sets (Lesson 1)

Note: Give the appropriate Practice Set to each student. Students who completed all questions correctly on their most recent Practice Set should be given the next level of difficulty. All other students should try to improve their scores on their current levels.

Students complete as many problems as they can in 90 seconds. Assign a counting pattern and start number for early finishers, or have them practice make ten addition or subtraction on the back of their papers. Collect and correct any Practice Sets completed within the allotted time.

Lesson 14: Add a pair of two-digit numbers when the ones digits have a sum greater than 10 using decomposition.

177

A STORY OF UNITS

Lesson 14 1•6

Add Tens (3 minutes)

Materials: (S) Personal white board, die per pair of students

Note: This fluency activity reviews adding multiples of 10 to two-digit numbers.

- Partner A writes or draws a number (with quick tens and ones) between 10 and 40 (e.g., 25).
- Partner B rolls the die to determine the number of tens to add (e.g., if he rolls 5, add 5 tens).
- Both partners write the number sentence on their personal white boards and check each other's work (e.g., 25 + 50 = 75).

Take Out Ones (5 minutes)

Materials: (S) Personal white board

Note: Taking out some ones from a two-digit number strengthens students' ability to apply the make ten strategy when adding two two-digit numbers.

Give students a sequence of related numbers at a time, and have them write number bonds on their personal white boards. Challenge early finishers to think of additional related number bonds for each sequence. Follow the suggested sequence:

- Take out 1: 8, 18, 28; 6, 56, 86.
- Take out 2: 5, 15, 25; 7, 37, 97.
- Take out 3: 6, 36, 76; 9, 69, 99, 109.
- Take out 4: 8, 48, 88, 108; 7, 77, 107, 117.

Concept Development (32 minutes)

Materials: (T) Chart paper, document camera if available
(S) Personal white board

Begin today's lesson with students at their desks or tables with their personal white boards.

Similar to the last two days, today's lesson provides opportunities for students to practice solving two-digit addition problems.

Today, however, in each set, a string of problems is related (e.g., 56 + 21, 56 + 24, and 56 + 27). For students who need additional support, the movement through the problems from simple to complex can help them choose a solution strategy.

Challenge students who are becoming proficient at solving two-digit addition problems to identify the relationship between each problem and create other strings that would exemplify the same set of relationships. Use their problems in the class if possible.

> **NOTES ON MULTIPLE MEANS OF ACTION AND EXPRESSION:**
>
> Students may choose how they want to solve problems—with drawings, number bonds, or the arrow way. Students should begin to move away from drawing to the more abstract methods of problem solving. However, not all students are ready, so support students wherever they are in their learning, and guide them as they progress.

178 Lesson 14: Add a pair of two-digit numbers when the ones digits have a sum greater than 10 using decomposition.

A STORY OF UNITS

Lesson 14 1•6

As in Lessons 12 and 13, invite students to share their methods for solving using place value language to explain why they chose to solve using these methods.

Problems 1–6 use easier combinations of ones as they create sums in the ones place that are equal to or greater than 10.

Problems 7–12 use combinations of ones that are typically more challenging for students.

Problems 1–6	**Problems 7–12**
65 + 15	56 + 28
65 + 16	46 + 28
65 + 19	38 + 56
48 + 33	37 + 57
48 + 43	37 + 47
38 + 62	45 + 37

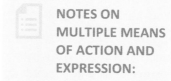

NOTES ON MULTIPLE MEANS OF ACTION AND EXPRESSION:

Continue to challenge students working above grade level. Change some of the expressions into number sentences with missing addends, or give students some word problems to solve with similar numbers.

Below are some of the various methods and explanations that students might share:

$$46 + 28 = 74$$
$$20 \quad 8$$
$$46 + 20 = 66$$
$$66 + 8 = 74$$
$$4 \quad 4$$

I added 20 to 46 first.

$$46 + 28 = 74$$
$$4 \quad 24$$
$$46 + 4 = 50$$
$$50 + 24 = 74$$
$$20 \quad 4$$

I made a ten first.

$$46 + 28 = 74$$
$$40 \quad 6 \quad 20 \quad 8$$
$$40 + 20 = 60$$
$$6 + 8 = 14$$
$$60 + 14 = 74$$

I added the 4 tens to 2 tens first.

Problem Set (10 minutes)

Students should do their personal best to complete the Problem Set within the allotted 10 minutes. For some classes, it may be appropriate to modify the assignment by specifying which problems they work on first. Some problems do not specify a method for solving. Students should solve these problems using the RDW approach used for Application Problems.

Lesson 14: Add a pair of two-digit numbers when the ones digits have a sum greater than 10 using decomposition.

A STORY OF UNITS — Lesson 14 1•6

Student Debrief (10 minutes)

Lesson Objective: Add a pair of two-digit numbers when the ones digits have a sum greater than 10 using decomposition.

The Student Debrief is intended to invite reflection and active processing of the total lesson experience.

Invite students to review their solutions for the Problem Set. They should check work by comparing answers with a partner before going over answers as a class. Look for misconceptions or misunderstandings that can be addressed in the Debrief. Guide students in a conversation to debrief the Problem Set and process the lesson.

Any combination of the questions below may be used to lead the discussion.

- Look at Problem 1 (a) and (b). How can solving Problem 1(a) help you solve Problem 1(b)?
- Look at Problem 2 (g) and (h). How are they related? How could solving one help you solve the other?
- Think about Take Out Ones in our Fluency Practice today. How did it help you when you were solving the more challenging problems?

Exit Ticket (3 minutes)

After the Student Debrief, instruct students to complete the Exit Ticket. A review of their work will help with assessing students' understanding of the concepts that were presented in today's lesson and planning more effectively for future lessons. The questions may be read aloud to the students.

Name: Maria   Date: _____

1. Solve and show your work.

   a. $48 + 21 = 69$
      $48 + 20 = 68$
      $68 + 1 = 69$

   b. $48 + 22 = 70$
      $48 + 2 = 50$
      $50 + 20 = 70$

   c. $39 + 43 = 82$
      $39 + 1 = 40$
      $40 + 42 = 82$

   d. $48 + 34 = 82$
      $48 + 30 = 78$
      $78 + 4 = 82$

   e. $77 + 14 = 91$
      $77 + 3 = 80$
      $80 + 11 = 91$

   f. $67 + 27 = 94$
      $67 + 3 = 70$
      $70 + 24 = 94$

   g. $58 + 37 = 95$
      $58 + 30 = 88$
      $88 + 7 = 95$

   h. $68 + 29 = 97$
      $68 + 2 = 70$
      $70 + 27 = 97$

2. Solve and show your work.

   a. $39 + 31 = 70$
      $39 + 1 = 40$
      $40 + 30 = 70$

   b. $58 + 23 = 81$
      $58 + 2 = 60$
      $60 + 21 = 81$

   c. $77 + 23 = 100$
      $77 + 3 = 80$
      $80 + 20 = 100$

   d. $69 + 26 = 95$
      $69 + 1 = 70$
      $70 + 25 = 95$

   e. $68 + 25 = 93$
      $68 + 20 = 88$
      $88 + 5 = 93$

   f. $45 + 37 = 82$
      $45 + 5 = 50$
      $50 + 32 = 82$

   g. $59 + 39 = 98$
      $50 + 30 = 80$
      $9 + 9 = 18$
      $80 + 18 = 98$

   h. $58 + 38 = 96$
      $50 + 30 = 80$
      $8 + 8 = 16$
      $80 + 16 = 96$

A STORY OF UNITS

Lesson 14 Problem Set 1•6

Name _____   Date _____

1. Solve and show your work.

a. 48 + 21 = _____	b. 48 + 22 = _____
c. 39 + 43 = _____	d. 48 + 34 = _____
e. 77 + 14 = _____	f. 67 + 27 = _____
g. 58 + 37 = _____	h. 68 + 29 = _____

Lesson 14: Add a pair of two-digit numbers when the ones digits have a sum greater than 10 using decomposition.

2. Solve and show your work.

a. 39 + 31 = _____	b. 58 + 23 = _____
c. 77 + 23 = _____	d. 69 + 26 = _____
e. 68 + 25 = _____	f. 45 + 37 = _____
g. 59 + 39 = _____	h. 58 + 38 = _____

A STORY OF UNITS

Lesson 14 Exit Ticket 1•6

Name _____ Date _____

Solve and show your work.

a. 47 + 42 = _____

b. 78 + 22 = _____

c. 56 + 38 = _____

Lesson 14: Add a pair of two-digit numbers when the ones digits have a sum greater than 10 using decomposition.

A STORY OF UNITS

Lesson 14 Homework 1•6

Name _____ Date _____

1. Solve and show your work.

a. 68 + 21 = _____	b. 59 + 32 = _____
c. 39 + 44 = _____	d. 58 + 36 = _____
e. 76 + 17 = _____	f. 68 + 26 = _____
g. 56 + 39 = _____	h. 58 + 29 = _____

Lesson 14: Add a pair of two-digit numbers when the ones digits have a sum greater than 10 using decomposition.

2. Solve and show your work.

a. 39 + 41 = ____

b. 48 + 43 = ____

c. 87 + 13 = ____

d. 59 + 25 = ____

e. 65 + 27 = ____

f. 27 + 67 = ____

g. 49 + 39 = ____

h. 38 + 58 = ____

A STORY OF UNITS

Lesson 15 1•6

Lesson 15

Objective: Add a pair of two-digit numbers when the ones digits have a sum greater than 10 with drawing. Record the total below.

Suggested Lesson Structure

- Application Problem   (5 minutes)
- Fluency Practice   (10 minutes)
- Concept Development   (35 minutes)
- Student Debrief   (10 minutes)

Total Time   **(60 minutes)**

Application Problem (5 minutes)

There are 20 students in class. Nine students put away their backpacks. How many more students still need to put away their backpacks?

Note: This is a *take apart with addend unknown* problem type that is set in a typical classroom context. Take note of students' independent choices to solve using addition or subtraction number sentences.

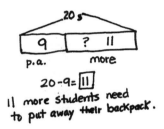

Fluency Practice (10 minutes)

- Core Fluency Differentiated Practice Sets **1.OA.6**   (5 minutes)
- Take Out Ones **1.OA.6, 1.NBT.4**   (5 minutes)

Core Fluency Differentiated Practice Sets (5 minutes)

Materials: (S) Core Fluency Practice Sets (Lesson 1)

Note: Give the appropriate Practice Set to each student. Help students become aware of their improvement. After students complete today's Practice Sets, ask them to stand if they tried a new level today or improved their scores from the previous day. Consider having students clap once for each person standing to celebrate improvement.

Students complete as many problems as they can in 90 seconds. Assign a counting pattern and start number for early finishers, or have them practice make ten addition or subtraction on the back of their papers. Collect and correct any Practice Sets completed within the allotted time.

186   Lesson 15:   Add a pair of two-digit numbers when the ones digits have a sum greater than 10 with drawing. Record the total below.

A STORY OF UNITS

Lesson 15 1•6

Take Out Ones (5 minutes)

Materials: (S) Personal white board

Note: Taking out some ones from a two-digit number strengthens students' ability to apply the make ten strategy when adding two two-digit numbers.

Repeat from the previous lesson. Give students a sequence of related numbers, and have them write number bonds on their personal white boards. Challenge early finishers to think of additional related number bonds for each sequence. Follow the suggested sequence:

- Take out 1: 2, 42, 72; 5, 55, 85.
- Take out 2: 7, 47, 67; 9, 69, 99.
- Take out 3: 8, 58, 78; 7, 67, 97, 107.
- Take out 4: 6, 46, 86, 106; 9, 79, 109, 119.

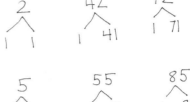

Concept Development (35 minutes)

Materials: (T) 10 ten-sticks (5 red, 5 yellow) (S) 5 ten-sticks, personal white board, place value chart (Lesson 3 Template 2)

Students sit in the meeting area with their materials in a semicircle formation.

- T: (Write 59 + 34 = ___.) I want to show this problem with the ten-sticks. What is the total number of tens in the first addend?
- S: 5 tens.
- T: (Project 5 ten-sticks onto the board.) We have 5 tens and how many more ones?
- S: 9 ones.
- T: (Project 9 cubes arranged in a 5-group formation, as shown to the right.)

- T: How many tens are in 34?
- S: 3 tens.
- T: Will we be adding 3 tens to the ones or to the tens?
- S: To the tens.
- T: (Vertically align 3 ten-sticks to the 5 ten-sticks.) 34 is 3 tens and how many more ones?
- S: 4 ones.
- T: We should add them to…?
- S: The ones!

NOTES ON MULTIPLE MEANS OF REPRESENTATION:

Support students who may have difficulty lining up their numbers to add vertically. These students may benefit from more concrete or pictorial supports while adding. Have them use the place value chart more regularly until they are able to line up the digits independently.

Lesson 15:   Add a pair of two-digit numbers when the ones digits have a sum greater than 10 with drawing. Record the total below.

187

A STORY OF UNITS Lesson 15 1•6

T: (Vertically align 4 ones to 9 ones as shown.) Our cubes are arranged, so we are ready to add. What is 9 ones and 4 ones? Turn and talk to your partner about what I can do with the ones.

S: 13 ones. → 9 needs 1 more to make ten. Take 1 from the 4. Now we have 10 and 3.

T: (Group the 9 and 1 cube on the board.) Now that we made a new ten, how many ones do we still have?

S: 3 ones.

T: (Write 3 in the ones place.) How many tens do we have now? Explain your thinking to your partner.

S: 9 tens. → 5 tens and 3 tens is 8 tens. We also made a new ten when we added 9 and 4, so that makes 9 tens altogether.

T: (Write 9 in the tens place.) So, what is 59 + 34? Say the number sentence.

S: 59 + 34 = 93.

Repeat the process using the following sequence:

- 49 + 35
- 43 + 36
- 38 + 47
- 17 + 65
- 38 + 52
- 38 + 62

Beginning at 17 + 65, have students make quick ten drawings to show their work.

T: (Write 17 + 65 = ___.) Make a quick ten drawing to show the first addend.

S: (Draw 1 quick ten and 7 ones.)

T: (Circulate and make sure the students arrange their 7 circles in 5-groups.)

T: Let's get ready to draw 65. Where should we draw the 6 quick tens?

S: Under the tens, right below the 1 ten from 17.

T: Where should we draw the 5 ones?

S: Under the ones, right below the 7 ones from 17.

T: Draw 65 and solve. (Circulate and support students as needed.)

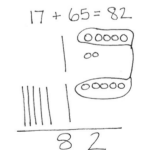

NOTES ON MULTIPLE MEANS OF REPRESENTATION:

Students demonstrate a true understanding of math concepts when they make connections and apply them in a variety of situations. By scaffolding questions, it is possible to guide connections, analysis, and mastery in students.

Problem Set (10 minutes)

Students should do their personal best to complete the Problem Set within the allotted 10 minutes. For some classes, it may be appropriate to modify the assignment by specifying which problems they work on first. Some problems do not specify a method for solving. Students should solve these problems using the RDW approach used for Application Problems.

188   Lesson 15: Add a pair of two-digit numbers when the ones digits have a sum greater than 10 with drawing. Record the total below.

A STORY OF UNITS

Lesson 15 1•6

Student Debrief (10 minutes)

Lesson Objective: Add a pair of two-digit numbers when the ones digits have a sum greater than 10 with drawing. Record the total below.

The Student Debrief is intended to invite reflection and active processing of the total lesson experience.

Invite students to review their solutions for the Problem Set. They should check work by comparing answers with a partner before going over answers as a class. Look for misconceptions or misunderstandings that can be addressed in the Debrief. Guide students in a conversation to debrief the Problem Set and process the lesson.

Any combination of the questions below may be used to lead the discussion.

- Look at Problem 2 (c), (d), and (e). How can they have the same answer but different numbers?
- Look at Problem 1 (b) or (d). Why is it more efficient to add the ones first instead of the tens?
- How does lining up the ones and tens help us with adding?
- How is lining up the ones and tens similar to and different from using the make ten strategy to add?
- Which is easier for you? Adding by lining up our ones and tens or using the number bonds? Explain your thinking.
- How did today's fluency activity help you solve today's addition problems?

Exit Ticket (3 minutes)

After the Student Debrief, instruct students to complete the Exit Ticket. A review of their work will help with assessing students' understanding of the concepts that were presented in today's lesson and planning more effectively for future lessons. The questions may be read aloud to the students.

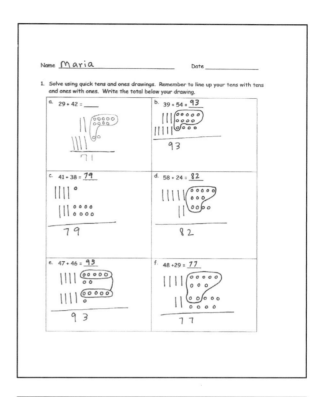

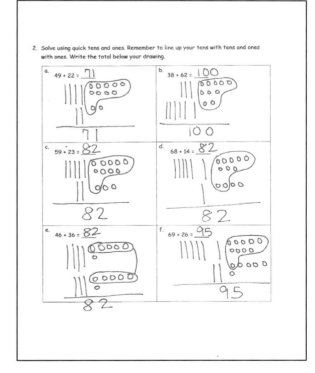

Lesson 15: Add a pair of two-digit numbers when the ones digits have a sum greater than 10 with drawing. Record the total below.

189

A STORY OF UNITS Lesson 15 Problem Set 1•6

Name _____ Date _____

1. Solve using quick tens and ones drawings. Remember to line up your tens with tens and ones with ones. Write the total below your drawing.

    a. 29 + 42 = ____

    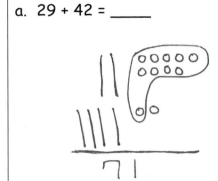

    71

    b. 39 + 54 = ____

    c. 41 + 38 = ____

    d. 58 + 24 = ____

    e. 47 + 46 = ____

    f. 48 + 29 = ____

Lesson 15: Add a pair of two-digit numbers when the ones digits have a sum greater than 10 with drawing. Record the total below.

Lesson 15 Problem Set 1•6

2. Solve using quick tens and ones. Remember to line up your tens with tens and ones with ones. Write the total below your drawing.

a. 49 + 22 = _____	b. 38 + 62 = _____
c. 59 + 23 = _____	d. 68 + 14 = _____
e. 46 + 36 = _____	f. 69 + 26 = _____

Lesson 15: Add a pair of two-digit numbers when the ones digits have a sum greater than 10 with drawing. Record the total below.

A STORY OF UNITS

Lesson 15 Exit Ticket 1•6

Name _____ Date _____

Solve using quick tens and ones drawings. Remember to line up your drawings and write the total below your drawing.

a. 49 + 34 = _____	b. 57 + 36 = _____

Lesson 15: Add a pair of two-digit numbers when the ones digits have a sum greater than 10 with drawing. Record the total below.

A STORY OF UNITS

Lesson 15 Homework 1•6

Name _____   Date _____

1. Solve using quick tens and ones drawings. Remember to line up your tens with tens and ones with ones. Write the total below your drawing.

a. 39 + 42 = _____	b. 48 + 36 = _____
c. 31 + 48 = _____	d. 47 + 34 = _____
e. 57 + 39 = _____	f. 58 + 27 = _____

Lesson 15: Add a pair of two-digit numbers when the ones digits have a sum greater than 10 with drawing. Record the total below.

193

2. Solve using quick tens and ones. Remember to line up your tens with tens and ones with ones. Write the total below your drawing.

a. 59 + 25 = _____

b. 48 + 42 = _____

c. 39 + 53 = _____

d. 78 + 14 = _____

e. 57 + 25 = _____

f. 69 + 27 = _____

Lesson 16

Objective: Add a pair of two-digit numbers when the ones digits have a sum greater than 10 with drawing. Record the new ten below.

Suggested Lesson Structure

- ■ Application Problem (5 minutes)
- ■ Fluency Practice (13 minutes)
- ■ Concept Development (32 minutes)
- ■ Student Debrief (10 minutes)
- **Total Time** **(60 minutes)**

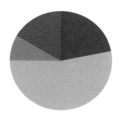

Application Problem (5 minutes)

Fifteen students ordered pizza for lunch. Seven students brought their lunch from home. How many fewer students brought their lunch from home than ordered lunch?

Note: Today's Application Problem is a *compare with difference unknown* problem type. Consider altering the meal choice to match the school's lunch menu for the day.

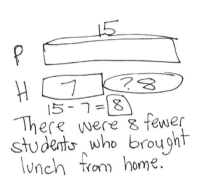

Fluency Practice (13 minutes)

- Grade 1 Core Fluency Sprint **1.OA.6**   (10 minutes)
- Coin Drop **1.NBT.5, 1.MD.3**   (3 minutes)

Grade 1 Core Fluency Sprint (10 minutes)

Materials: (S) Core Fluency Sprints (Lesson 3)

Note: Choose a Sprint based on the needs of the class.

- Core Addition Sprint 1
- Core Addition Sprint 2
- Core Subtraction Sprint
- Core Fluency Sprint: Totals of 5, 6, and 7
- Core Fluency Sprint: Totals of 8, 9, and 10

Lesson 16: Add a pair of two-digit numbers when the ones digits have a sum greater than 10 with drawing. Record the new ten below.

195

A STORY OF UNITS

Lesson 16   1•6

Coin Drop (3 minutes)

Materials: (T) 4 dimes, 10 pennies, can

Note: In this activity, students practice adding and subtracting ones and tens within 100.

- T: (Hold up a penny.) Name my coin.
- S: A penny.
- T: How much is it worth?
- S: 1 cent.
- T: Listen carefully as I drop coins in my can. Count along in your minds.

> **NOTES ON MULTIPLE MEANS OF ENGAGEMENT:**
>
> After playing Coin Drop with pennies and then dimes, mix pennies and dimes so that students have to add based on the changing value of the coin. This challenges students and keeps them listening for what is coming.

Drop in some pennies, and ask how much money is in the can. Take out some pennies, and show them. Ask how much money is still in the can. Continue adding and subtracting pennies for a minute. Then, repeat the activity with dimes. For the final minute, begin with some pennies in the can, and add and subtract dimes.

Concept Development (32 minutes)

Materials: (T) Chart paper (S) Personal white board, recording tens and ones (Template)

Gather students in the meeting area with their materials.

- T: (Write 39 + 26 = ___ on the board.) On your personal white board, make a quick ten drawing to solve.
- S: (Solve as the teacher circulates and selects one student to share her solution with the class.)
- T: (Choose a student, Student 1, to model the drawing on the board.) As Student 1 draws and explains what she did, I'm going to stop her after every step to show how we can record using just numbers.
- S1: (Draws 39.) I drew 3 tens and 9 ones.
- T: Stop! She made 3 tens, so I write 3 in the tens place. She made 9 ones, so I write the 9 in the…?
- S: Ones place!
- T: (Write 39.)
- S1: (Draws 26 vertically aligned to 39.) I drew 2 tens and 6 ones right below so I can add tens to tens and ones to ones.
- T: Stop! Watch me as I match exactly what Student 1 did with her drawing. (Write 26.) I'm adding the 2 tens to the 3 tens, 6 ones to the 9 ones, just like the picture. (Draw the equal sign.)
- S1: Then, I added the ones together. 9 needs 1 from 6 to get to 10. (Frames 10.) 10 and 5 is 15.

> **NOTES ON MULTIPLE MEANS OF ACTION AND EXPRESSION:**
>
> Giving students an opportunity to share their thinking allows students to evaluate their process and practice. English language learners also benefit from hearing other students explain their thinking.

---

196   Lesson 16:   Add a pair of two-digit numbers when the ones digits have a sum greater than 10 with drawing. Record the new ten below.

©2015 Great Minds. eureka-math.org
G1-M6-TE-BK6-1.3.1-1.2016

A STORY OF UNITS   Lesson 16 1•6

T: Stop! Student 1 made 15 by adding 9 and 6. (Point to the digits in the ones place.) That's 1 ten 5 ones. Watch where I record that new ten. (Record the new ten below the second addend in the tens place as shown to the right.) I didn't write the 1 ten where the answer goes yet because I have more tens to add later. 15 is 1 ten and…?

S: 5 ones.

T: (Write 5 in the ones place.)

S1: Then, I added 3 tens and 2 tens plus the 1 ten I made when I added 9 and 6. That's 6 tens. (Writes 6 in the tens place.)

T: Ah, ha! So, she added 3 tens and 2 tens (point to the digits 3 and 2) plus this new ten we wrote in from 15 when we added the ones. So 3 tens + 2 tens + 1 ten is…?

S: 6 tens.

T: So, what is 39 + 26? Say the number sentence.

S: 39 + 26 = 65.

T: Let's try some more.

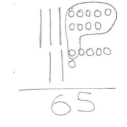

NOTES ON MULTIPLE MEANS OF ACTION AND EXPRESSION:

Continue to challenge students working above grade level. Encourage students to write a creative word problem to match one of the number sentences they solved today.

Continue with the following process using the suggested sequence as the students are ready:

39 + 36, 59 + 37, 28 + 43, 47 + 35, 26 + 67.

- Have another student model the quick ten drawings as the teacher represents the drawings with numbers.
- The teacher draws the quick ten drawings, and students represent the drawings with just numbers on the place value chart.
- Students make the quick ten drawings *and* represent them with just numbers side by side.

MP.4

Problem Set (10 minutes)

Students should do their personal best to complete the Problem Set within the allotted 10 minutes. For some classes, it may be appropriate to modify the assignment by specifying which problems they work on first. Some problems do not specify a method for solving. Students should solve these problems using the RDW approach used for Application Problems.

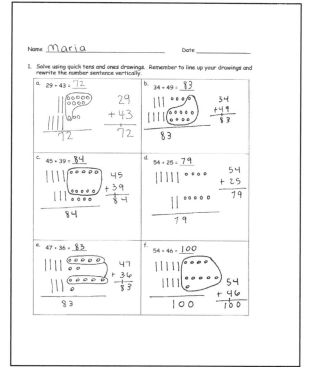

Lesson 16: Add a pair of two-digit numbers when the ones digits have a sum greater than 10 with drawing. Record the new ten below.

Lesson 16

Student Debrief (10 minutes)

Lesson Objective: Add a pair of two-digit numbers when the ones digits have a sum greater than 10 with drawing. Record the new ten below.

The Student Debrief is intended to invite reflection and active processing of the total lesson experience.

Invite students to review their solutions for the Problem Set. They should check work by comparing answers with a partner before going over answers as a class. Look for misconceptions or misunderstandings that can be addressed in the Debrief. Guide students in a conversation to debrief the Problem Set and process the lesson.

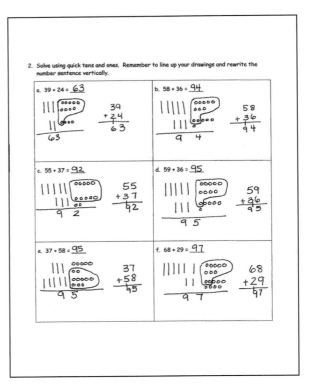

Any combination of the questions below may be used to lead the discussion.

- Look at page 1 of your Problem Set. What is different about Problem 1(d) compared to the others?
- Look at Problem 1(f). Why is there a zero in the ones place in the answer when we added some ones together in the problem?
- What new math notation did we use today to communicate how we added precisely?
- Do you prefer to add by lining up your tens and ones or by using the number bond to add?

Exit Ticket (3 minutes)

After the Student Debrief, instruct students to complete the Exit Ticket. A review of their work will help with assessing students' understanding of the concepts that were presented in today's lesson and planning more effectively for future lessons. The questions may be read aloud to the students.

A STORY OF UNITS　　　　　　　　　　　　　　　　Lesson 16 Problem Set　1•6

Name _____   Date _____

1. Solve using quick tens and ones drawings. Remember to line up your drawings and rewrite the number sentence vertically.

a. 29 + 43 = ____

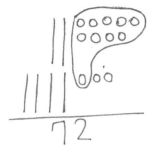

```
  29
+ 43
-----
  72
```
72

b. 34 + 49 = ____

c. 45 + 39 = ____

d. 54 + 25 = ____

e. 47 + 36 = ____

f. 54 + 46 = ____

Lesson 16: Add a pair of two-digit numbers when the ones digits have a sum greater than 10 with drawing. Record the new ten below.

199

A STORY OF UNITS

Lesson 16 Problem Set 1•6

2. Solve using quick tens and ones. Remember to line up your drawings and rewrite the number sentence vertically.

a. 39 + 24 = _____

b. 58 + 36 = _____

c. 55 + 37 = _____

d. 59 + 36 = _____

e. 37 + 58 = _____

f. 68 + 29 = _____

Lesson 16: Add a pair of two-digit numbers when the ones digits have a sum greater than 10 with drawing. Record the new ten below.

A STORY OF UNITS

Lesson 16 Exit Ticket 1•6

Name _____ Date _____

Solve using quick tens and ones. Remember to line up your drawings and rewrite the number sentence vertically.

a. 49 + 26 = _____	b. 58 + 37 = _____
c. 55 + 37 = _____	d. 69 + 26 = _____

Lesson 16: Add a pair of two-digit numbers when the ones digits have a sum greater than 10 with drawing. Record the new ten below.

201

A STORY OF UNITS

Lesson 16 Homework 1•6

Name _____ Date _____

1. Solve using quick tens and ones drawings. Remember to line up your drawings and rewrite the number sentence vertically.

    29
   +43

   72

a. 39 + 45 = ____	b. 64 + 28 = ____
c. 47 + 38 = ____	d. 53 + 27 = ____
e. 38 + 48 = ____	f. 53 + 45 = ____

Lesson 16: Add a pair of two-digit numbers when the ones digits have a sum greater than 10 with drawing. Record the new ten below.

Lesson 16 Homework 1•6

2. Solve using quick tens and ones. Remember to line up your drawings and rewrite the number sentence vertically.

a. 79 + 14 = ____

b. 28 + 47 = ____

c. 58 + 33 = ____

d. 19 + 66 = ____

e. 39 + 59 = ____

f. 49 + 48 = ____

Tens	Ones

recording tens and ones

Lesson 17

Objective: Add a pair of two-digit numbers when the ones digits have a sum greater than 10 with drawing. Record the new ten below.

Suggested Lesson Structure

- ■ Application Problem   (5 minutes)
- ■ Fluency Practice   (13 minutes)
- ■ Concept Development   (32 minutes)
- ■ Student Debrief   (10 minutes)
- **Total Time   (60 minutes)**

Application Problem (5 minutes)

Rose saw 14 monkeys at the zoo. She saw 5 fewer monkeys than foxes. How many foxes did Rose see?

Note: Today's problem is a *comparison with larger unknown* problem type where *fewer* suggests the wrong operation. Students should be exposed to these problems, but mastery is not expected until the end of Grade 2.

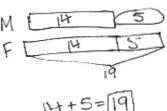

Fluency Practice (13 minutes)

- Grade 1 Core Fluency Sprint **1.OA.6**   (10 minutes)
- Analogous Addition Sentences **1.OA.6, 1.NBT.4**   (3 minutes)

Grade 1 Core Fluency Sprint (10 minutes)

Materials: (S) Core Fluency Sprints (Lesson 3)

Note: Based on the needs of the class, select a Sprint from yesterday's materials. There are several possible options available.

1. Re-administer the Sprint from the day before.
2. Administer the next Sprint in the sequence.
3. Differentiate. Administer two different Sprints. Simply have one group do a counting activity on the back of the Sprint while the other Sprint is corrected.

A STORY OF UNITS

Lesson 17 1•6

Analogous Addition Sentences (3 minutes)

Note: This fluency activity encourages students to use sums within 10 to solve more challenging problems.

T: Say the number sentence with the answer. 5 + 2.
S: 5 + 2 = 7.
T: 45 + 2.
S: 45 + 2 = 47.
T: 42 + 5.
S: 42 + 5 = 47.
T: 5 + 42.
S: 5 + 42 = 47.

Continue with the following suggested sequence:

4 + 3	6 + 3	5 + 4
84 + 3	76 + 3	95 + 4
83 + 4	73 + 6	94 + 5
4 + 83	6 + 73	5 + 94

Concept Development (32 minutes)

Materials: (T) Chart paper (S) Personal white board, recording tens and ones (Lesson 16 Template) (optional), numeral cards (Lesson 3 Fluency Template)

Students sit at their tables with their personal white boards.

This Concept Development can be used to solidify the learning acquired in Lessons 15 and 16. Three sets of problems have been provided for students who are ready to extend their double-digit addition skills. The teaching sequence from Lesson 16 may be used to guide instruction. Students should be encouraged to solve by using quick ten drawings as well as the standard algorithm. Encourage students to use place value language to describe strategies for solving.

Problems 1–4	Problems 5–8	Problems 9–12
25 + 13	49 + 25	55 + 39
29 + 13	58 + 32	36 + 57
39 + 23	67 + 28	15 + 78
38 + 25	67 + 26	27 + 73

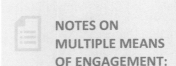

NOTES ON MULTIPLE MEANS OF ENGAGEMENT:

Appropriate scaffolds help all students feel successful. As students work, observe closely to determine if any would benefit from one-on-one problem-solving assistance.

If time allows, have student pairs use numeral cards to generate two-digit addition problems to solve with their partners. This gives the teacher an opportunity to work in a small group with students who need extra support.

Lesson 17: Add a pair of two-digit numbers when the ones digits have a sum greater than 10 with drawing. Record the new ten below.

A STORY OF UNITS

Lesson 17 1•6

- Create a tens pile (digits 0–4) and a ones pile (digits 5–9) using numeral cards from both players, and put them facedown.
- Put the place value chart templates between the partners, and add an addition sign in between the charts.
- Partner A creates the first addend by drawing a card from the tens pile and ones pile and places them in the first place value chart.
- Partner B creates the second addend in the same way, placing them in the second place value chart.
- Each student solves the problem with a quick ten drawing and the standard algorithm on his personal white board.

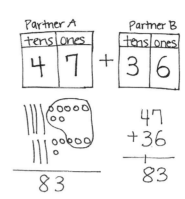

Problem Set (10 minutes)

Students should do their personal best to complete the Problem Set within the allotted 10 minutes. For some classes, it may be appropriate to modify the assignment by specifying which problems they work on first. Some problems do not specify a method for solving. Students should solve these problems using the RDW approach used for Application Problems.

NOTES ON MULTIPLE MEANS OF ACTION AND EXPRESSION:

Continue to challenge students who are working above grade level. After they have completed the Problem Set, encourage them to write a word problem adding a pair of two-digit numbers. Have students who write a word problem trade papers to solve each other's problem.

Student Debrief (10 minutes)

Lesson Objective: Add a pair of two-digit numbers when the ones digits have a sum greater than 10 with drawing. Record the new ten below.

The Student Debrief is intended to invite reflection and active processing of the total lesson experience.

Invite students to review their solutions for the Problem Set. They should check work by comparing answers with a partner before going over answers as a class. Look for misconceptions or misunderstandings that can be addressed in the Debrief. Guide students in a conversation to debrief the Problem Set and process the lesson.

Any combination of the questions below may be used to lead the discussion.

- How can solving Problem 1(b) help you solve Problem 1(c)?

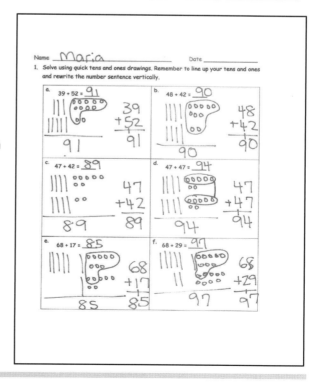

Lesson 17: Add a pair of two-digit numbers when the ones digits have a sum greater than 10 with drawing. Record the new ten below.

207

A STORY OF UNITS   Lesson 17 1•6

- Look at your quick ten drawing for Problem 2(d). How did you make a new ten? Show another way to make a new ten.
- Look at Problems 1(a) and 2(a) with a partner. How are these problems related? How can solving Problem 1(a) help you solve Problem 2(a)? Think of another problem you could solve that is related to Problems 1(a) and 2(a).
- Look at Problem 1 (c) and (d). How are these problems alike? Why is the total of 47 and 42 a number in the 80's and the total of 47 and 47 is a number in the 90's?
- Which addition strategy do you prefer? Explain your thinking.
- How did the Analogous Addition Sentences help you with addition during today's lesson?

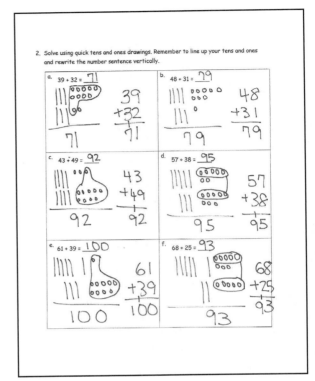

Exit Ticket (3 minutes)

After the Student Debrief, instruct students to complete the Exit Ticket. A review of their work will help with assessing students' understanding of the concepts that were presented in today's lesson and planning more effectively for future lessons. The questions may be read aloud to the students.

Lesson 17: Add a pair of two-digit numbers when the ones digits have a sum greater than 10 with drawing. Record the new ten below.

A STORY OF UNITS

Lesson 17 Problem Set 1•6

Name _____ Date _____

1. Solve using quick tens and ones drawings. Remember to line up your tens and ones and rewrite the number sentence vertically.

a. 39 + 52 = ____	b. 48 + 42 = ____
c. 47 + 42 = ____	d. 47 + 47 = ____
e. 68 + 17 = ____	f. 68 + 29 = ____

Lesson 17: Add a pair of two-digit numbers when the ones digits have a sum greater than 10 with drawing. Record the new ten below.

209

A STORY OF UNITS

Lesson 17 Problem Set 1•6

2. Solve using quick tens and ones drawings. Remember to line up your tens and ones and rewrite the number sentence vertically.

a. 39 + 32 = _____	b. 48 + 31 = _____
c. 43 + 49 = _____	d. 57 + 38 = _____
e. 61 + 39 = _____	f. 68 + 25 = _____

Lesson 17: Add a pair of two-digit numbers when the ones digits have a sum greater than 10 with drawing. Record the new ten below.

A STORY OF UNITS

Lesson 17 Exit Ticket 1•6

Name _____ Date _____

Solve using quick tens and ones drawings. Remember to line up your tens and ones and rewrite the number sentence vertically.

a. 39 + 47 = ____	b. 58 + 32 = ____
c. 49 + 44 = ____	d. 58 + 39 = ____

Lesson 17: Add a pair of two-digit numbers when the ones digits have a sum greater than 10 with drawing. Record the new ten below.

211

A STORY OF UNITS

Lesson 17 Homework 1•6

Name _____ Date _____

1. Solve using quick tens and ones drawings. Remember to line up your tens and ones and rewrite the number sentence vertically.

   a. 49 + 33 = _____

   b. 68 + 32 = _____

   c. 36 + 43 = _____

   d. 27 + 67 = _____

   e. 78 + 17 = _____

   f. 69 + 28 = _____

Lesson 17: Add a pair of two-digit numbers when the ones digits have a sum greater than 10 with drawing. Record the new ten below.

A STORY OF UNITS

Lesson 17 Homework 1•6

2. Solve using quick tens and ones drawings. Remember to line up your tens and ones and rewrite the number sentence vertically.

a. 29 + 52 = _____	b. 58 + 31 = _____
c. 73 + 26 = _____	d. 67 + 28 = _____
e. 41 + 59 = _____	f. 48 + 45 = _____

Lesson 17: Add a pair of two-digit numbers when the ones digits have a sum greater than 10 with drawing. Record the new ten below.

A STORY OF UNITS

1 GRADE

Mathematics Curriculum

GRADE 1 • MODULE 6

Topic D
Varied Place Value Strategies for Addition to 100

1.NBT.4

Focus Standard:	1.NBT.4	Add within 100, including adding a two-digit number and a one-digit number, and adding a two-digit number and a multiple of 10, using concrete models or drawings and strategies based on place value, properties of operations, and/or the relationship between addition and subtraction; relate the strategy to a written method and explain the reasoning used. Understand that in adding two-digit numbers, one adds tens and tens, ones and ones; and sometimes it is necessary to compose a ten.
Instructional Days:	2	
Coherence -Links from:	G1–M4	Place Value, Comparison, Addition and Subtraction to 40
-Links to:	G2–M3	Place Value, Counting, and Comparison of Numbers to 1,000

During Topic D, students discuss and compare the various place value strategies they use when adding to 100 (**1.NBT.4**). Students have the opportunity to explain their thinking and better understand the strategies based on the examples and explanations of peers.

Lesson 18 has students adding a pair of two-digit numbers, such as 36 + 57, in more than one way, explaining the similarities and differences in the methods. Students recognize that they can achieve the same accurate sum through the varied strategies, as they decompose and recompose the numbers, attending to the tens and ones.

Students share their preferred strategies in Lesson 19, explaining the reason they choose to use a particular strategy for a particular set of addends. For instance, when adding 39 + 43, one student may prefer to use the make ten strategy, decomposing 43 into 1 and 42, because adding 40 + 42 is an easy problem for her. Another student may prefer vertically aligning the numbers to ensure that he is adding ones with ones and then tens with tens. Students discuss questions such as, "In which number bonds do you see an easier problem to solve? Is there another way to solve this problem? How are [the selected student's] methods different from or the same as your partner's? What is a compliment you would like to give [him or her]?"

214   Topic D: Varied Place Value Strategies for Addition to 100

A STORY OF UNITS

Topic D 1•6

A Teaching Sequence Toward Mastery of Varied Place Value Strategies for Addition to 100

Objective 1: Add a pair of two-digit numbers with varied sums in the ones, and compare the results of different recording methods.
(Lesson 18)

Objective 2: Solve and share strategies for adding two-digit numbers with varied sums.
(Lesson 19)

Topic D: Varied Place Value Strategies for Addition to 100

Lesson 18

Objective: Add a pair of two-digit numbers with varied sums in the ones, and compare the results of different recording methods.

Suggested Lesson Structure

■ Fluency Practice   (13 minutes)
■ Application Problem   (5 minutes)
■ Concept Development   (32 minutes)
■ Student Debrief   (10 minutes)
  Total Time   **(60 minutes)**

Fluency Practice (13 minutes)

- Standards Check: Commutative Property **1.OA.3, 1.OA.7**   (5 minutes)
- Standards Check: Subtraction as Unknown Addend **1.OA.4**   (8 minutes)

Standards Check: Commutative Property (5 minutes)

Materials: (S) Pair of dice, personal white board

Note: In the remaining lessons, there are a variety of fluency activities that can be used to monitor students' mastery of grade level standards. Take note of any students who may need additional support or particular standards-based activities that may be useful to include in summer practice.

This activity reviews the commutative property of addition (e.g., if 6 + 3 = 9 is known, then 3 + 6 = 9 is also known) (**1.OA.3**) and requires students to understand the meaning of the equal sign (**1.OA.7**).

$$6+3=9$$
$$3+6=9$$
$$9=6+3$$
$$9=3+6$$

- Assign partners.
- Both partners roll a die and then write four addition sentences using the rolled numbers as addends.
- Partners check each other's work.

Standards Check: Subtraction as Unknown Addend (8 minutes)

Materials: (S) Pattern sheet list A or B (Fluency Template)

Note: This activity provides review with converting subtraction expressions to unknown addend equations.

- Assign partners of equal ability, and give one partner List A and the other List B.

A STORY OF UNITS

Lesson 18 1•6

- Students convert the subtraction expressions on their lists to addition equations with unknown addends (e.g., for 10 – 9, the student would write 9 + ___ = 10).
- Partners exchange lists and solve.

Application Problem (5 minutes)

A farmer counted 12 bunnies in their cages in the morning. In the afternoon, he only counted 4 bunnies in their cages. How many bunnies disappeared from their cages?

Note: Today's problem is a *take away with change unknown* problem type. As Topic F, which focuses on varied problem types, approaches, begin to take note of students' strengths and weaknesses for specific problem types.

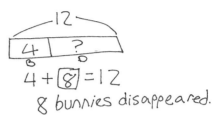

8 bunnies disappeared.

Concept Development (32 minutes)

Materials: (T) Student work samples (Template), projector (S) Personal white board

[Student work samples showing Student A and Student B solving 58 + 37 = 95 using number bonds]

Have students sit at their tables or in the meeting area with their personal boards.

T: (Write 58 + 37 on the board.) Solve this problem. (Pause while students work. Quietly post a second problem for early finishers.)
T: The answer is...?
S: 95.
T: Take a moment to discuss your strategy and/or correct your work with your partner.
T: (Project work from Student A and Student B.) Let's compare Student A's work to Student B's work. What is the same, and what is different about their solution strategies? Turn and talk to your partner.
S: They both used number bonds. → Both students broke apart 37. → They both used tens to solve.
T: I have two labels. Read them to me.
S: Make the Next Ten. Count On by Tens First.

Lesson 18: Add a pair of two-digit numbers with varied sums in the ones, and compare the results of different recording methods.

217

A STORY OF UNITS Lesson 18 1•6

T: Talk to your partner. Which label best describes the solution strategy of each student? Explain why.

S: Student A made the next ten first. → Student A broke 37 into 2 and 35 so he could add 2 and 58 to make 60. → Student B counted on by tens. That's why he broke apart 37 into 30 and 7. 58 and 30 is 88. → Student B added the tens first. I don't think he counted on by tens, but I guess that label fits the best.

T: (Label Student A's work *Make the Next Ten*. Label Student B's work *Count On by Tens First*.)

T: Can both students' work be correct even though they used tens in different ways?

S: Yes!

T: What is a compliment you can give to each of these students?

S: They drew correct number bonds. → Student A showed how she made the next ten from 58. You can see that in the number bond and in the first addition sentence. → Student B did a good job by breaking apart the tens from 37 so he could add 3 tens to 58 first.

MP.3

T: What are some ways they could improve their work?

S: Student B could write an addition sentence that showed how he got 88. But maybe he did that in his head.

> **NOTES ON MULTIPLE MEANS OF REPRESENTATION:**
>
> Facilitate student discussions to provide opportunities for comprehension. Guide students to recognize strategies that can make math easier, for example, breaking a larger number into number bonds as well as looking for patterns and structures in their work.

T: (Project Student C's work.) How did Student C solve 58 + 37? Turn and talk to your partner.

S: He drew quick tens and ones by lining up the tens to tens and the ones to ones. → Then, he showed exactly how he added using just the numbers.

T: (Label the work *The Quick Ten Drawing—Adding Tens to Tens and Ones to Ones*.)

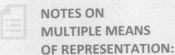

T: This student's answer is 85, instead of 95 like we got. What happened? Can we find the error in his work?

S: When he added the ones together, he made the next ten with 8 and 2 from the 7. But when he added the tens, he forgot about the next ten! → You can see that when he used just the numbers. He didn't remember the next ten. It's easier to remember a next ten when you write it in the tens place. → There should be a total of 9 tens, not 8 tens. The answer is 9 tens 5 ones. 95.

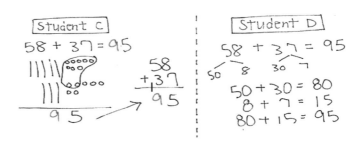

T: Discuss with your partner. What are some ways this student can improve his work?

S: He can work more carefully and realize that he made the next ten. → He can record the next ten. Then, he can catch his mistake. → The student can look at his picture to check his work.

218 Lesson 18: Add a pair of two-digit numbers with varied sums in the ones, and compare the results of different recording methods.

A STORY OF UNITS

Lesson 18

T: Yes! It is important to record when you have made the next ten. It helps to keep track of all of your thinking.

T: Rewrite this student's work on your board, solving it correctly. When you're finished, check your work with your partner.

S: (Work with partners to solve using quick ten drawings.)

T: (As students finish, choose a pair of students to show their work on the board as the new work for Student C.)

T: (Project Student D's work.) Let's compare Student D's work to Student C's new work. What similarities and differences do you notice? Turn and talk to your partner.

S: They look different because Student D used number bonds and three addition sentences to solve the problem. But our new work for Student C shows quick ten drawings with lined up numbers to add tens with tens and ones with ones. → They both added ones to ones and then tens to tens! They both added 8 and 7 and got 15. Then, they added 5 tens and 3 tens to get 8 tens. Then, they both added the next ten and got 95.

T: (Write 47 + 36 on the board.) Solve a new problem. You may use any method to solve, but you must show your work.

Have students swap boards with their partner and discuss the following:

- How did your partner show her solution?
- How was her work different from your work?
- How was your work the same?
- Give your partner a compliment on her work.
- Give a suggestion for how she could improve her work.

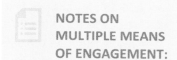

NOTES ON MULTIPLE MEANS OF ENGAGEMENT:

Appropriate scaffolds help all students feel successful. As students are working, observe closely to determine if any would benefit from one-on-one problem-solving assistance.

Project three or four work samples from the class, showing each of the methods: vertical alignment, make the next ten, count on by tens, and add ones to ones and then tens to tens using number bonds.

T: How is the student work shown different from your partner's work?

If time allows, have students solve 26 + 65, and then share another set of student work from the class.

Problem Set (10 minutes)

Students should do their personal best to complete the Problem Set within the allotted 10 minutes. For some classes, it may be appropriate to modify the assignment by specifying which problems they work on first. Some problems do not specify a method for solving. Students should solve these problems using the RDW approach used for Application Problems.

Lesson 18: Add a pair of two-digit numbers with varied sums in the ones, and compare the results of different recording methods.

A STORY OF UNITS

Lesson 18   1•6

Student Debrief (10 minutes)

Lesson Objective: Add a pair of two-digit numbers with varied sums in the ones, and compare the results of different recording methods.

The Student Debrief is intended to invite reflection and active processing of the total lesson experience.

Invite students to review their solutions for the Problem Set. They should check work by comparing answers with a partner before going over answers as a class. Look for misconceptions or misunderstandings that can be addressed in the Debrief. Guide students in a conversation to debrief the Problem Set and process the lesson.

Any combination of the questions below may be used to lead the discussion.

- Look at Problem 2. Which strategy, count on by tens first or make the next ten first, would you use to solve? Explain your choice.
- Why didn't most of us use the make the next ten strategy when solving Problem 1?
- The make the next ten strategy and another strategy, too, can be used for Problem 1. Explain to your partner why these number sentences are correct. (Write 74 + 21 = 80 + 15, 74 + 21 = 65 + 30, and 74 + 21 = 75 + 20.)
- How can solving Problem 5 help you solve Problem 6?
- Which strategy do you find yourself using the most? Why do you prefer that strategy?

Exit Ticket (3 minutes)

After the Student Debrief, instruct students to complete the Exit Ticket. A review of their work will help with assessing students' understanding of the concepts that were presented in today's lesson and planning more effectively for future lessons. The questions may be read aloud to the students.

Lesson 18: Add a pair of two-digit numbers with varied sums in the ones, and compare the results of different recording methods.

A STORY OF UNITS

Lesson 18 Problem Set 1•6

Name _____ Date _____

Use any method you prefer to solve the problems below.

1. 74 + 21 = _____	2. 79 + 21 = _____
3. 46 + 34 = _____	4. 58 + 34 = _____
5. 35 + 14 = _____	6. 35 + 18 = _____

Lesson 18: Add a pair of two-digit numbers with varied sums in the ones, and compare the results of different recording methods.

A STORY OF UNITS

Lesson 18 Exit Ticket 1•6

Name _____ Date _____

Circle the work that is correct.

In the extra space, correct the mistake in the other solution using the same solution strategy the student tried to use.

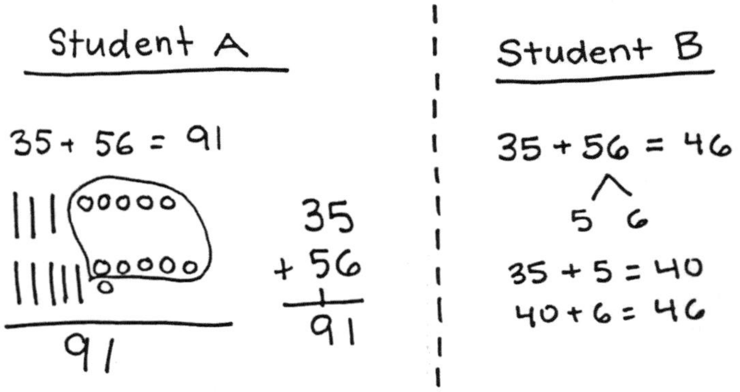

Student A

35 + 56 = 91

|||(ooooo)   35
|||||(ooooo) +56
      o   ___
              91
     91

Student B

35 + 56 = 46
      ∧
     5 6
35 + 5 = 40
40 + 6 = 46

A STORY OF UNITS

Lesson 18 Homework 1•6

Name _____ Date _____

Use any method you prefer to solve the problems below.

1. 61 + 15 = _____	2. 16 + 51 = _____
3. 37 + 45 = _____	4. 27 + 46 = _____
5. 58 + 27 = _____	6. 38 + 48 = _____

Lesson 18: Add a pair of two-digit numbers with varied sums in the ones, and compare the results of different recording methods.

A STORY OF UNITS Lesson 18 Fluency Template 1•6

Name _____

Partner _____

   Example

Step 1: Rewrite 4 − 1 as 1 + ___ = 4.

Step 2: Exchange papers and solve.

List A

1. 10 − 9 _____
2. 10 − 8 _____
3. 9 − 8 _____
4. 9 − 6 _____
5. 8 − 6 _____
6. 7 − 4 _____
7. 7 − 5 _____
8. 8 − 5 _____
9. 9 − 5 _____
10. 9 − 6 _____

Name _____

Partner _____

   Example

Step 1: Rewrite 4 − 1 as 1 + ___ = 4.

Step 2: Exchange papers and solve.

List B

1. 10 − 8 _____
2. 10 − 7 _____
3. 8 − 7 _____
4. 8 − 6 _____
5. 9 − 6 _____
6. 7 − 6 _____
7. 7 − 5 _____
8. 7 − 4 _____
9. 8 − 5 _____
10. 6 − 4 _____

pattern sheet list A or B

Lesson 18: Add a pair of two-digit numbers with varied sums in the ones, and compare the results of different recording methods.

EUREKA MATH

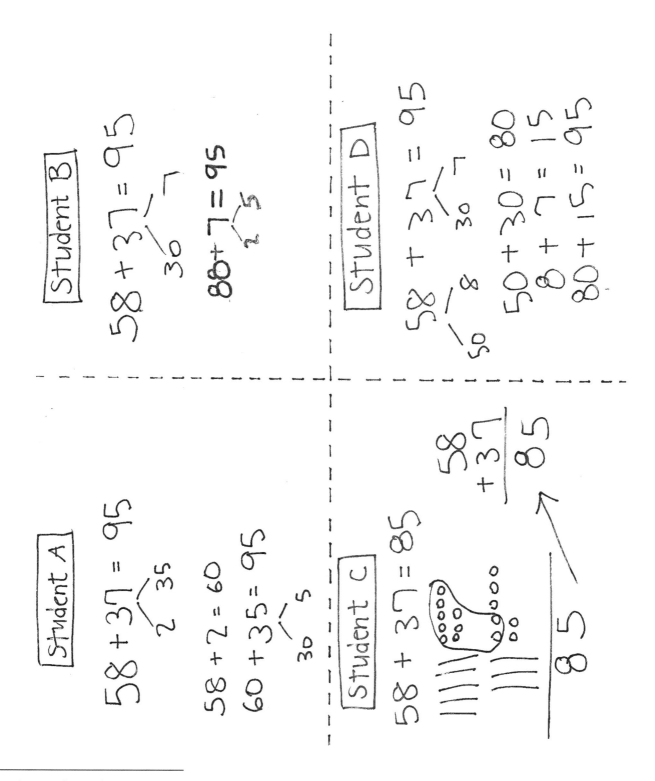

student work samples

Lesson 18: Add a pair of two-digit numbers with varied sums in the ones, and compare the results of different recording methods.

Lesson 19

Objective: Solve and share strategies for adding two-digit numbers with varied sums.

Suggested Lesson Structure

- ■ Fluency Practice   (13 minutes)
- ■ Application Problem   (5 minutes)
- ■ Concept Development   (32 minutes)
- ■ Student Debrief   (10 minutes)
- **Total Time**   **(60 minutes)**

Fluency Practice (13 minutes)

- Core Fluency Differentiated Practice Sets **1.OA.6**   (5 minutes)
- Standards Check: True or False Number Sentences **1.OA.7**   (8 minutes)

Core Fluency Differentiated Practice Sets (5 minutes)

Materials: (S) Core Fluency Practice Sets (Lesson 1)

Note: Give the appropriate Practice Set to each student. Help students become aware of their improvement. After students do today's Practice Sets, ask them to stand if they tried a new level today or improved their score from the previous day. Consider having students clap for each person standing to celebrate improvement.

Students complete as many problems as they can in 90 seconds. Assign a counting pattern and start number for early finishers, or have them practice make ten addition or subtraction on the back of their papers. Collect and correct any Practice Sets completed within the allotted time.

Standards Check: True or False Number Sentences (8 minutes)

Materials: (S) Personal white board

Note: Use professional judgment to determine whether students would benefit more from repeating the previous standards check or moving on to this one. Today's standards check reviews the meaning of the equal sign and requires students to determine if addition and subtraction equations are true or false.

T: (Write 5 = 1 + 4.) What's 1 + 4?
S: 5

A STORY OF UNITS   Lesson 19 1•6

T: (Write 5 = 5 directly underneath 5 = 1 + 4.) Is 5 = 1 + 4 true or false?
S: True.
T: Why?
S: Because 5 is equal to 5. → Because 5 is the same as 5.
T: Now, you do the same. Rename the side of the number sentence with a plus or minus symbol as one number.
T: (Write 7 = 3 + 5.)
S: (Write 7 = 8.)
T: Show me your boards. (Pause to see.) Is 7 = 3 + 5 true or false?
S: False.
T: Why?
S: Because 7 is not the same as 8. → Because 7 doesn't equal 8.
T/S: (Draw a line through the equal sign to show 7 ≠ 3 + 5 and 7 ≠ 8 to record they are not true.)

As time permits, continue with the following suggested sequence:

a. 7 = 2 + 5   d. 7 − 2 = 4   g. 6 + 1 = 5 + 2   j. 8 − 5 = 9 − 4
b. 3 + 6 = 9   e. 3 = 8 − 5   h. 4 + 3 = 7 + 1   k. 8 − 6 = 2 + 4
c. 8 = 2 + 7   f. 3 = 9 − 7   i. 8 − 4 = 6 − 2   l. 4 + 5 = 9 − 3

Application Problem (5 minutes)

Ben had 16 baseball cards before a card show. After the card show, he had 20 baseball cards. How many cards were added to Ben's collection?

Note: Today's problem is an *add to with change unknown* problem type. As Topic F, which focuses on varied problem types, approaches, begin to take note of students' strengths and weaknesses for specific problem types.

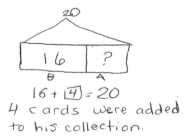

Concept Development (32 minutes)

Materials: (T) Projector (S) Personal white board

Students sit at their tables next to their partners with personal boards.

T: Solve 39 + 43 using any strategy we've learned so far. Be ready to explain why you chose the strategy. (Circulate and note the types of strategies being used.)
S: (Solve.)

> **NOTES ON MULTIPLE MEANS OF ACTION AND EXPRESSION:**
>
> Giving students an opportunity to share their thinking allows them to evaluate their process and practice. English language learners also benefit from hearing others explain their thinking.

Lesson 19: Solve and share strategies for adding two-digit numbers with varied sums.

227

T: Turn and talk to your partner, and share your work. Explain to your partner why you chose that particular strategy. What similarities and differences do you notice between your work and your partner's?

S: (Explain and compare strategies.)

T: (While student pairs share their work, ask two or three students to come up and write their work on the board. Be sure to include students who solved using different strategies.)

T: Let's hear how our friends solved 39 + 43 and why they chose to use their particular strategy.

S: (Make the next ten strategy.) I know that 39 is really close to 40, so I took 1 from 43. I saw it as 40 + 42. That's 82.

(Vertical alignment.) It's quick and easy for me to add 9 and 3 and 3 tens and 4 tens. I can see which digits I need to add more clearly when I line up the tens to tens and ones to ones.

(Standard algorithm.) I can line up my tens and ones without using drawings.

(Adding on tens first.) I am really good at adding tens onto any number. 39 and 40 is 79. Then, I added 3 to get 82.

(Compensation.) I thought of it a different way, like a balance. 39 + 43. Add one to 39 and subtract one from 43, so it's 40 + 42.

MP.5

As each student explains the work and shares the reasons for his or her strategy choice, have students discuss questions such as the ones listed below:

- Is there another way to solve this problem?
- How does the number bond make it easier to add the parts?
- How is Student A's strategy different or the same as your partner's?
- When do you think is the best time to use the make ten strategy?
- What compliment can you give him?
- What advice can you give him to make the work better?
- Repeat the process possibly using the following suggested sequence:

    66 + 29

    56 + 35

    18 + 78

    34 + 47

> **NOTES ON MULTIPLE MEANS OF ACTION AND EXPRESSION:**
>
> Continue to challenge students working above grade level. After they have completed the Problem Set, encourage them to write a word problem to match one of the number sentences. Have students who write word problems trade papers and try to find which number sentence the word problem matches.

Problem Set (10 minutes)

Students should do their personal best to complete the Problem Set within the allotted 10 minutes. For some classes, it may be appropriate to modify the assignment by specifying which problems they work on first. Some problems do not specify a method for solving. Students should solve these problems using the RDW approach used for Application Problems.

A STORY OF UNITS

Lesson 19 1•6

Student Debrief (10 minutes)

Lesson Objective: Solve and share strategies for adding two-digit numbers with varied sums.

The Student Debrief is intended to invite reflection and active processing of the total lesson experience.

Invite students to review their solutions for the Problem Set. They should check work by comparing answers with a partner before going over answers as a class. Look for misconceptions or misunderstandings that can be addressed in the Debrief. Guide students in a conversation to debrief the Problem Set and process the lesson.

Any combination of the questions below may be used to lead the discussion.

- How can solving Problem 1 help you solve Problem 2?
- Explain how Problems 3 and 4 are related. Can you see that they would have the same sum without calculating the sum?
- Which strategy do you use the most? Why? Do you study the numbers and choose a specific strategy that works better with those numbers, or do you always use the same strategy? Use an example from your Problem Set to explain your reasoning.
- Today, we changed our number sentences to be very simple. We changed 5 + 3 = 7 to 8 = 7. We changed 4 = 3 + 1 to 4 = 4. How did that help you see if the number sentences were true or false?

Exit Ticket (3 minutes)

After the Student Debrief, instruct students to complete the Exit Ticket. A review of their work will help with assessing students' understanding of the concepts that were presented in today's lesson and planning more effectively for future lessons. The questions may be read aloud to the students.

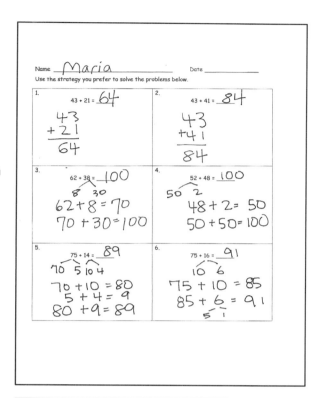

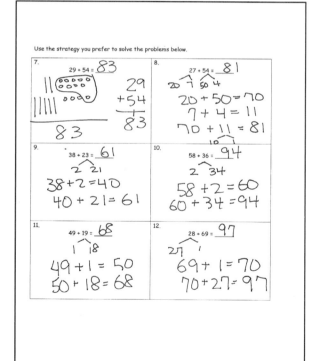

Lesson 19: Solve and share strategies for adding two-digit numbers with varied sums.

A STORY OF UNITS Lesson 19 Problem Set 1•6

Name _____ Date _____

Use the strategy you prefer to solve the problems below.

1. 43 + 21 = _____	2. 43 + 41 = _____
3. 62 + 38 = _____	4. 52 + 48 = _____
5. 75 + 14 = _____	6. 75 + 16 = _____

Lesson 19: Solve and share strategies for adding two-digit numbers with varied sums.

Use the strategy you prefer to solve the problems below.

7. 29 + 54 = _____

8. 27 + 54 = _____

9. 38 + 23 = _____

10. 58 + 36 = _____

11. 49 + 19 = _____

12. 28 + 69 = _____

A STORY OF UNITS

Lesson 19 Exit Ticket 1•6

Name _____ Date _____

Use the strategy you prefer to solve the problems below.

a.
  24 + 38 = _____

b.
  24 + 48 = _____

Lesson 19: Solve and share strategies for adding two-digit numbers with varied sums.

A STORY OF UNITS **Lesson 19 Homework 1•6**

Name _____ Date _____

Use the strategy you prefer to solve the problems below.

1. 53 + 22 = _____	2. 23 + 52 = _____
3. 76 + 14 = _____	4. 76 + 16 = _____
5. 55 + 35 = _____	6. 54 + 46 = _____

Lesson 19: Solve and share strategies for adding two-digit numbers with varied sums.

A STORY OF UNITS **Lesson 19 Homework 1•6**

Use the strategy you prefer to solve the problems below.

7. 49 + 25 = _____	8. 49 + 45 = _____
9. 37 + 37 = _____	10. 37 + 57 = _____
11. 24 + 48 = _____	12. 26 + 68 = _____

Name _____ Date _____

1. Use the RDW process to solve the following problems. Write your statement on the line.

   a. Lucy has 5 pencils. Kim has 7 pencils. How many more pencils does Kim have than Lucy?

   _____.

   b. Ben has 18 pencils. Anton has 9 pencils. How many fewer pencils does Anton have than Ben?

   _____.

   c. Julio has 5 more pencils than Fran. Fran has 6 pencils. How many pencils does Julio have?

   _____.

A STORY OF UNITS

Mid-Module Assessment Task 1•6

2. Fill in the missing numbers in the sequence.

a.

97, 98, ____, ____, ____, ____

b.

116, 117, ____, ____, ____

c.

____, 14, ____, ____, 11, ____

d.

112, 111, ___, 109, ___, ___

3. Write the number as tens and ones in the place value chart, or use the place value chart to write the number.

a. 82

tens	ones

b. 99

tens	ones

c. _____

tens	ones
9	6

d. _____

tens	ones
10	5

4. Match the equal amounts.

a. 51   8 tens 6 ones

b. 68   8 ones 6 tens

c. 114   4 tens 11 ones

d. 86   11 tens 4 ones

236   Module 6:   Place Value, Comparison, Addition and Subtraction to 100

5. Use <, =, or > to compare the pairs of numbers.

   a. 69 ◯ 79

   b. 15 ◯ 50

   c. 99 ◯ 101

   d. 110 ◯ 108

   e. 61 ◯ 5 tens 11 ones

6. Ben thinks 92 ones is greater than 9 tens 2 ones. Is he correct? Explain your thinking using words, pictures, or numbers. Draw and write about tens and ones to explain your thinking.

7. Find the mystery numbers. Explain how you know the answers.

   a. 10 more than 90 is _____.

tens	ones
9	0
   →
tens	ones

   b. 10 less than 90 is _____.

tens	ones
9	0
   →
tens	ones

   c. 1 more than 90 is _____.

tens	ones
9	0
   →
tens	ones

   d. 1 less than 90 is _____.

tens	ones
9	0
   →
tens	ones

Module 6: Place Value, Comparison, Addition and Subtraction to 100

8. Solve for each unknown number. Use the space provided to show your work.

a. 80 + 6 = _____	b. 20 + _____ = 80
c. 7 tens – _____ = 4 tens	d. 90 – 40 = _____
e. 68 + 7 = _____	f. 51 + 20 = _____
g. 46 + 31 = _____	h. 46 + 35 = _____

Mid-Module Assessment Task

Topics A–D

Standards Addressed

Represent and solve problems involving addition and subtraction.

1.OA.1 Use addition and subtraction within 20 to solve word problems involving situations of adding to, taking from, putting together, taking apart, and comparing, with unknowns in all positions, e.g., by using objects, drawings, and equations with a symbol for the unknown number to represent the problem. (See CCSS-M Glossary, Table 1.)

Extend the counting sequence.

1.NBT.1 Count to 120, starting at any number less than 120. In this range, read and write numerals and represent a number of objects with a written numeral.

Understand place value.

1.NBT.2 Understand that the two digits of a two-digit number represent amounts of tens and ones. Understand the following as special cases:

 a. 10 can be thought of as a bundle of ten ones—called a "ten."

 c. The numbers 10, 20, 30, 40, 50, 60, 70, 80, 90 refer to one, two, three, four, five, six, seven, eight, or nine tens (and 0 ones).

1.NBT.3 Compare two two-digit numbers based on meanings of the tens and ones digits, recording the results of comparisons with the symbols >, =, and <.

Use place value understanding and properties of operations to add and subtract.

1.NBT.4 Add within 100, including adding a two-digit number and a one-digit number, and adding a two-digit number and a multiple of 10, using concrete models or drawings and strategies based on place value, properties of operations, and/or the relationship between addition and subtraction; relate the strategy to a written method and explain the reasoning used. Understand that in adding two-digit numbers, one adds tens and tens, ones and ones; and sometimes it is necessary to compose a ten.

1.NBT.5 Given a two-digit number, mentally find 10 more or 10 less than the number, without having to count; explain the reasoning used.

1.NBT.6 Subtract multiples of 10 in the range 10–90 from multiples of 10 in the range 10–90 (positive or zero differences), using concrete modules or drawings and strategies based on place value, properties of operations, and/or the relationship between addition and subtraction; relate the strategy to a written method and explain the reasoning used.

Module 6: Place Value, Comparison, Addition and Subtraction to 100

Evaluating Student Learning Outcomes

A Progression Toward Mastery is provided to describe steps that illuminate the gradually increasing understandings that students develop *on their way to proficiency*. In this chart, this progress is presented from left (Step 1) to right (Step 4). The learning goal for students is to achieve Step 4 mastery. These steps are meant to help teachers and students identify and celebrate what the students CAN do now and what they need to work on next.

A Progression Toward Mastery

Assessment Task Item	STEP 1 Little evidence of reasoning without a correct answer. (1 Point)	STEP 2 Evidence of some reasoning without a correct answer. (2 Points)	STEP 3 Evidence of some reasoning with a correct answer or evidence of solid reasoning with an incorrect answer. (3 Points)	STEP 4 Evidence of solid reasoning with a correct answer. (4 Points)

1
 	
1.OA.1
	Student answers are incorrect, and there is no evidence of reasoning.	Student answers are incorrect, but there is evidence of reasoning. For example, student is able to write a number sentence.	Student answers are correct, but the responses are incomplete (e.g., may be missing labels for the drawing, an addition sentence, or an explanation). Student work is essentially strong.	Student correctly:
				• Solves each word problem.
				  a. Kim has 2 more pencils than Lucy.
				  b. Anton has 9 fewer pencils than Ben.
				  c. Julio has 11 pencils.
				• Demonstrates understanding of the problem situation through drawing/modeling.

2 1.NBT.1	Student is unable to complete any one sequence of numbers.	Student completes at least one sequence.	Student completes at least one sequence as well as at least two numbers in each additional sequence. OR Student completes two or more sequences correctly.	Student identifies all numbers in the sequences: • 97, 98, **99**, **100**, **101, 102** • 116, 117, **118, 119, 120** • **15**, 14, **13, 12,** 11, **10** • 112, 111, **110**, 109, **108, 107**

A STORY OF UNITS

Mid-Module Assessment Task 1•6

A Progression Toward Mastery				
3 1.NBT.2	Student does not demonstrate understanding of tens and ones and is unable to complete more than one answer correctly.	Student demonstrates inconsistent understanding of tens and ones, completing only two answers correctly.	Student demonstrates some understanding of most aspects of tens and ones, completing at least three answers correctly.	Student completes all correctly: a. 8-2 (or 7-12 or 0-82) b. 9-9 (or 0-99) c. 96 d. 105
4 1.NBT.2	Student does not demonstrate understanding of the equivalent representations of tens and ones and is unable to match any equal amounts.	Student demonstrates limited understanding of the equivalent representations of tens and ones, matching one or two equal amounts.	Student demonstrates some understanding of the equivalent representations of tens and ones, matching three equal amounts.	Student matches all four equal amounts as follows: a. 51 = **4 tens 11 ones** b. 68 = **8 ones 6 tens** c. 114 = **11 tens 4 ones** d. 86 = **8 tens 6 ones**
5 1.NBT.3	Student is unable to use symbols to compare numbers and is unable to correctly answer more than one of the five comparisons.	Student has limited ability to use symbols to compare numbers, correctly answering two of the five comparisons.	Student has some ability to use symbols to compare numbers, correctly answering three or four of the five comparisons.	Student correctly answers: a. < b. < c. < d. > e. =
6 1.NBT.2	Student demonstrates little to no understanding of comparing numbers based on tens and ones, answering incorrectly. There is no evidence of reasoning.	Student uses drawings or words to accurately depict at least one of the two numbers, demonstrating limited understanding of the use of place value to compare numbers.	Student demonstrates some understanding of using place value to compare numbers and correctly identifies the greater number but does not fully explain reasoning using place value. OR Student answers incorrectly because of an error such as transcription but demonstrates strong understanding of place value through drawing or words.	Student correctly uses drawings or words that depict place value to accurately explain that 92 ones is the same as 9 tens 2 ones.

Module 6: Place Value, Comparison, Addition and Subtraction to 100

A STORY OF UNITS
Mid-Module Assessment Task 1•6

A Progression Toward Mastery				
7 **1.NBT.5** **1.NBT.2**	Student demonstrates little or no understanding of mentally adding or subtracting 10. Answers are incorrect, and there is no evidence of reasoning.	Student demonstrates limited understanding of mentally adding or subtracting 10, identifying at least two correct mystery numbers, but does not complete any charts accurately.	Student demonstrates the ability to mentally add or subtract 10, correctly identifying four mystery numbers, but reasoning is unclear because no charts have been completed accurately. OR Student accurately completes charts but makes an error in mental calculation on one or two of (a), (b), (c), or (d).	Student identifies the following: a. 100 b. 80 c. 91 d. 89 and accurately completes the charts to depict the arrow way.

8
 	
1.NBT.4
1.NBT.6
	Student demonstrates little or no ability to add or subtract two-digit numbers to 100, answering two or fewer questions correctly.	Student demonstrates some ability to add (or subtract) two-digit numbers, answering at least four of eight correctly, and demonstrates misunderstandings of place value.	Student demonstrates the ability to add (and subtract) two-digit numbers, answering at least six of eight correctly, or uses a sound process throughout with four calculation errors at most.	Student correctly:
				• Solves
				  a. 86
				  b. 60
				  c. 3 tens
				  d. 50
				  e. 75
				  f. 71
				  g. 77
				  h. 81
				• Represents the process to accurately solve through drawings, number bonds, or the arrow way. The notation demonstrates the use of a sound strategy for adding or subtracting.

Module 6: Place Value, Comparison, Addition and Subtraction to 100

EUREKA MATH

A STORY OF UNITS   Mid-Module Assessment Task   1•6

Name __Maria__   Date _____

1. Use the RDW process to solve the following problems. Write your statement on the line.

   a. Lucy has 5 pencils. Kim has 7 pencils. How many more pencils does Kim have than Lucy?

   $7 - 5 = \boxed{2}$

   __Kim has 2 more pencils than Lucy.__

   b. Ben has 18 pencils. Anton has 9 pencils. How many fewer pencils does Anton have than Ben?

   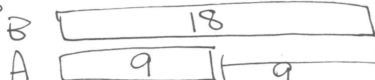

   $9 + \boxed{9} = 18$

   __Anton has 9 fewer pencils than Ben.__

   c. Julio has 5 more pencils than Fran. Fran has 6 pencils. How many pencils does Julio have?

   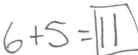
   $6 + 5 = \boxed{11}$

   __Julio has 11 pencils.__

Module 6:   Place Value, Comparison, Addition and Subtraction to 100   243

2. Fill in the missing numbers in the sequence.

a. 97, 98, __99__, __100__, __101__, __102__

b. 116, 117, __118__, __119__, __120__

c. __15__, 14, __13__, __12__, 11, __10__

d. 112, 111, __110__, 109, __108__, __107__

3. Write the number as tens and ones in the place value chart, or use the place value chart to write the number.

a. 82

tens	ones
8	2

b. 99

tens	ones
9	9

c. __96__

tens	ones
9	6

d. __105__

tens	ones
10	5

4. Match the equal amounts.

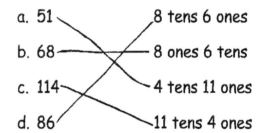

a. 51 — 4 tens 11 ones
b. 68 — 8 ones 6 tens
c. 114 — 11 tens 4 ones
d. 86 — 8 tens 6 ones

Mid-Module Assessment Task 1•6

5. Use <, =, or > to compare the pairs of numbers.

   a. 69 < 79

   b. 15 < 50

   c. 99 < 101

   d. 110 > 108

   e. 61 = 5 tens 11 ones

6. Ben thinks 92 ones is greater than 9 tens 2 ones. Is he correct? Explain your thinking using words, pictures or numbers. Draw and write about tens and ones to explain your thinking. 92 ones is the same as 9 tens 2 ones. 90 ones is 9 tens so 90+2 is the same as 90+2
10+10+10+10+10+10+10+10+10=90   92=92

7. Find the mystery numbers. Explain how you know the answers.

   a. 10 more than 90 is 100

tens	ones		tens	ones
9	0	→	10	0

   b. 10 less than 90 is 80

tens	ones		tens	ones
9	0	→	8	0

   c. 1 more than 90 is 91

tens	ones		tens	ones
9	0	→	9	1

   d. 1 less than 90 is 89

tens	ones		tens	ones
9	0	→	8	9

Module 6: Place Value, Comparison, Addition and Subtraction to 100

8. Solve for each unknown number. Use the space provided to show your work.

a. 80 + 6 = __86__ 80 + 6 ―― 86	b. 20 + __60__ = 80 ‖‖ ‖‖‖‖‖‖
c. 7 tens − __3 tens__ = 4 tens ‖‖‖‖ ⟨‖‖‖⟩ (crossed out)	d. 90 − 40 = __50__ 9 tens − 4 tens = 5 tens
e. 68 + 7 = __75__ ∧ 2 5	f. 51 + 20 = __71__ ∧ 50 1
g. 46 + 31 = __77__ ∧ 30 1	h. 46 + 35 = __81__ 46 + 35 ―― 81

A STORY OF UNITS

GRADE 1

Mathematics Curriculum

GRADE 1 • MODULE 6

Topic E
Coins and Their Values

1.MD.3

Focus Standard:	1.MD.3[1]	Tell and write time in hours and half-hours using analog and digital clocks. Recognize and identify coins, their names, and their value.
Instructional Days:	5	
Coherence -Links from:	G1–M4	Place Value, Comparison, Addition and Subtraction to 40
-Links to:	G2–M3	Place Value, Counting, and Comparison of Numbers to 1,000

Through Topic E, students learn about the four most predominant U.S. coins in circulation: the penny, the nickel, the dime, and the quarter. Students identify and use the coins based on their image, name, or value (**1.MD.3**).

In Lesson 20, students are introduced to the nickel, which they then use alongside

the familiar dime and penny. Students consider various ways to represent common values. For instance, students represent a value of 10 by using 1 ten (the dime) or 10 ones (pennies), as well as the well-known decomposition of 5 + 5 (2 nickels). Students use their background with number bonds to decompose the larger value into the various compositions.

Lesson 21 introduces students to the quarter, which can be the most challenging coin to learn. Students build on their understanding from Lesson 20, focusing specifically on the value of 25. They consider how many pennies they would need to have the same value as 1 quarter and then trade in 2 dimes and 1 nickel or 2 dimes and 5 pennies for a quarter. Again, students use their prior work with number bonds and place value charts to consider the various compositions.

During Lesson 22, students continue to work with all four coins. Various sequences are provided to best match the learning needs of the class. Finally, in Lesson 23, students count on from any coin to create various values.

To culminate the topic, students use dimes and pennies as representations of numbers to 120, connecting the prior knowledge students have developed throughout the module to their work in Topic E.

---

[1]Focus on money.

Topic E: Coins and Their Values

247

©2015 Great Minds. eureka-math.org
G1-M6-TE-BK6-1.3.1-1.2016

Topic E

A Teaching Sequence Toward Mastery of Coins and Their Values

Objective 1: Identify pennies, nickels, and dimes by their image, name, or value. Decompose the values of nickels and dimes using pennies and nickels.
(Lesson 20)

Objective 2: Identify quarters by their image, name, or value. Decompose the value of a quarter using pennies, nickels, and dimes.
(Lesson 21)

Objective 3: Identify varied coins by their image, name, or value. Add one cent to the value of any coin.
(Lesson 22)

Objective 4: Count on using pennies from any single coin.
(Lesson 23)

Objective 5: Use dimes and pennies as representations of numbers to 120.
(Lesson 24)

Lesson 20

Objective: Identify pennies, nickels, and dimes by their image, name, or value. Decompose the values of nickels and dimes using pennies and nickels.

Suggested Lesson Structure

- ■ Fluency Practice   (15 minutes)
- ■ Application Problem   (5 minutes)
- ■ Concept Development   (30 minutes)
- ■ Student Debrief   (10 minutes)
- **Total Time**   **(60 minutes)**

Fluency Practice (15 minutes)

- Grade 1 Core Fluency Sprint **1.OA.6**   (10 minutes)
- Standards Check: True or False Number Sentences **1.NBT.3**   (5 minutes)

Grade 1 Core Fluency Sprint (10 minutes)

Materials: (S) Core Fluency Sprints (Lesson 3)

Note: Choose a Sprint based on the needs of the class.

- Core Addition Sprint 1
- Core Addition Sprint 2
- Core Subtraction Sprint
- Core Fluency Sprint: Totals of 5, 6, and 7
- Core Fluency Sprint: Totals of 8, 9, and 10

Standards Check: True or False Number Sentences (5 minutes)

Materials: (S) Personal white board

Write a true or false number sentence. Students write a happy face on their personal boards if the number sentence is true. If the sentence is false, students write it with the correct symbol. Notice which problem types are difficult for them.

A STORY OF UNITS

Lesson 20 1•6

Use the first two columns (a–h) as the suggested sequence. At each checkpoint, decide whether students are ready for the next column or whether they should continue with similar problem types. The third column (i–l) is provided as a possible opportunity for a few students who would really enjoy a challenge.

a. 5 > 4
b. 50 > 40
c. 57 > 75
d. 16 < 51
Checkpoint.

e. 40 + 5 = 45
f. 73 = 7 + 30
g. 82 < 8 tens 2 ones
h. 97 > 9 ones 7 tens
Checkpoint.

i. 9 + 7 = 10 + 6
j. 16 + 10 = 26 – 10
k. 12 – 6 > 9
l. 90 < 89 + 1

Application Problem (5 minutes)

Tamra saw 10 cheetahs at the zoo. She saw 8 more leopards than cheetahs. How many leopards did she see?

Note: Today's problem is a *compare with bigger unknown* problem type. Some students may incorrectly solve the problem because of their reliance on the term *more*, rather than on their understanding of the comparison. Look at students' drawings to see how they made meaning of the problem.

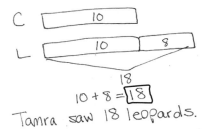

Concept Development (30 minutes)

Materials: (T) 5 dimes, 15 pennies, and 3 nickels (plastic or real) (S) 5 dimes, 15 pennies, 3 nickels (plastic or real), personal white board, spinner (Template) (optional), paper clip, pencil

Gather students in the meeting area with their materials.

T: (Lay out or project 1 dime.) What is the name of this coin?
S: A dime.
T: What is the value of one dime?
S: 10 cents!
T: Take out your dime, and show it to me. (Wait as students take it out. On chart paper, record the dime using a circle with the number 10 in it.)
T: I want a number of pennies to equal the value of a dime. How many pennies do I need?
S: 10 pennies!
T: Why do I need 10 pennies to have 1 dime?
S: Pennies are worth 1 cent. You need 10 pennies to make 10 cents. → A dime is worth the same as 10 pennies.
T: So, 1 dime (point to the dime on the chart paper) is equal to 10 pennies. Count the pennies for me as I draw, and when we get to 10, don't say 10 pennies but...
S: 1 dime!
T: Count as I point.
S: 1 penny, 2 pennies, 3 pennies, ..., 9 pennies, 1 dime.

---

250 | Lesson 20: Identify pennies, nickels, and dimes by their image, name, or value. Decompose the values of nickels and dimes using pennies and nickels.

A STORY OF UNITS — Lesson 20 1•6

T: (Hold up or project a nickel.) Two of these together have the same value as a dime. (Create a number bond with the coins, as shown to the right. Record the number bond, leaving out the value of the nickels.)

T: What is the value of this coin? Turn and talk with a partner, and make a number bond to show your thinking. Tell your partner how you know. (Wait as students discuss.)

T: What is the value?

S: 5 cents! (After students show their boards, add the value 5 to the two number bond parts.)

T: How do you know?

S: The number bond needs the same number for both parts. So, it must be 2 fives to make 10. → It's like a doubles fact. 5 + 5 = 10, so they must be five cents each. → I have nickels at home. I know they are worth 5 cents.

T: This coin is called a **nickel**. Find all the nickels in your bag. (Wait as students identify the nickels.)

T: Sort the rest of your coins into piles, so we can easily get what we need for today's lesson. Put each pile on your personal white board, and write the name and value of the coin under the pile. (Wait as students sort dimes, pennies, and nickels.)

T: What is one of the ways we made 10 cents?

S: We made 10 cents with 10 pennies. → We made 10 cents with 1 dime. → We made 10 cents with 2 nickels. → We made 10 cents with a nickel (5 cents) and 5 pennies (1 cent each).

T: (Display 2 nickels.) Two nickels is 10 cents. How many cents will I have when I put down 1 more nickel? (Wait as students determine the answer. Have them turn and talk as necessary.)

S: 15 cents!

T: Work with a partner to make 15 cents in different ways. (Wait and listen as students lay out coins to make 15 cents.)

T: How did you make 15 cents?

S: We used 15 pennies. Pennies are worth 1 cent. 15 pennies is 15 cents. → We used 1 dime and 5 pennies. That's 1 ten and 5 ones. → We did like you did. We got 3 nickels. → We used 1 dime and 1 nickel to make 15, since it's 10 and 5. → We used 2 nickels and 5 pennies. The two nickels make 10, and then 5 more pennies makes 15. (As student share, record their combinations on the chart paper.)

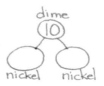

NOTES ON MULTIPLE MEANS OF REPRESENTATION:

Make sure both parts are the same number in students' number bonds. Since they are both the same kind of coin, the two parts must be the same value.

NOTES ON MULTIPLE MEANS OF REPRESENTATION:

If students struggle to generate combinations of coins, guide students through trading pennies for a nickel or a dime using questions such as, "How many pennies would we need to trade for a nickel? Do we have enough to do this?"

Lesson 20: Identify pennies, nickels, and dimes by their image, name, or value. Decompose the values of nickels and dimes using pennies and nickels.

251

Use the following suggested sequence, asking students to work with a partner to create a coin combination that has the given value. Record the combinations for each value on chart paper.

- 6 cents
- 11 cents
- 16 cents
- 20 cents

After students have successfully shown ways to make the above totals, provide the following riddles.

- T: (Project or write 2 + 3.) I want to use 1 coin to represent the total of 2 + 3. Which coin would I use? Tell a partner.
- T: Which coin could represent the total of 2 + 3?
- S: A nickel!
- T: How do you know?
- S: 2 + 3 = 5. → A nickel has a value of 5 cents.

Repeat the process with the following examples:

- 1 coin to represent the total of 6 + 4
- 1 coin to represent the total of 5 + 1 + 4
- 1 coin to represent the total of 1 + 0 or the value of 6 – 5
- 1 coin to represent the total of 4 + 1
- 2 coins to represent the total of 17 + 3
- 2 coins to represent the total of 2 + 8

If time permits, partners may play Coin Trade. The object of the game is to continue to trade coins, always having 10 cents.

Materials: Each player has 10 pennies, the spinner with a paper clip and pencil; each pair has a pot with pennies, nickels, and dimes for trading per pair.

- Partner A spins the spinner.
- Partner A trades pennies for the coin landed on. (For instance, if the student lands on a nickel, he trades 5 pennies for 1 nickel. If he lands on a dime, he trades 10 pennies for 1 dime. If he lands on a penny, he trades a penny for a penny.) Player A counts his coins to be sure he still has 10 cents.
- Partner B takes a turn. Player B counts her coins to be sure she still has 10 cents.
- Play continues as time allows.
- The person with the most pennies at the end of the game is the winner.

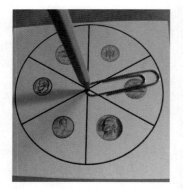

As play continues, students might land on the coins they already have, such as landing on a penny when they have 10 pennies. Students may trade one of their pennies for a new penny. Play the game for about five minutes or as time allows.

Lesson 20: Identify pennies, nickels, and dimes by their image, name, or value. Decompose the values of nickels and dimes using pennies and nickels.

Lesson 20

Problem Set (10 minutes)

Students should do their personal best to complete the Problem Set within the allotted 10 minutes. For some classes, it may be appropriate to modify the assignment by specifying which problems they work on first. Some problems do not specify a method for solving. Students should solve these problems using the RDW approach used for Application Problems.

Student Debrief (10 minutes)

Lesson Objective: Identify pennies, nickels, and dimes by their image, name, or value. Decompose the values of nickels and dimes using pennies and nickels.

The Student Debrief is intended to invite reflection and active processing of the total lesson experience.

Invite students to review their solutions for the Problem Set. They should check work by comparing answers with a partner before going over answers as a class. Look for misconceptions or misunderstandings that can be addressed in the Debrief. Guide students in a conversation to debrief the Problem Set and process the lesson. Any combination of the questions below may be used to lead the discussion.

- Look at Problem 1. What parts of the picture of each coin help you identify it?
- Look at Problem 4. Share your solutions. Are there only two ways to make 10 cents with your coins? How many different ways can we make 10 cents using our coins?
- If you had to carry around 10 cents all day, which combination of coins would you want to carry? Why?
- Which coin was new to us today? (**Nickel.**) Describe the coin in as many ways as you can.

Exit Ticket (3 minutes)

After the Student Debrief, instruct students to complete the Exit Ticket. A review of their work will help with assessing the students' understanding of the concepts that were presented in today's lesson and planning more effectively for future lessons. The questions may be read aloud to the students.

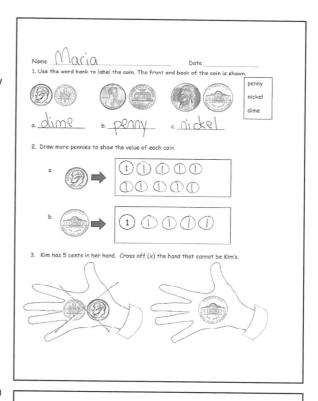

Lesson 20: Identify pennies, nickels, and dimes by their image, name, or value. Decompose the values of nickels and dimes using pennies and nickels.

253

Name _____ Date _____

1. Use the word bank to label the coin. The front and back of the coin is shown.

| penny |
| nickel |
| dime |

a. _____   b. _____   c. _____

2. Draw more pennies to show the value of each coin.

a.  ➡ ①

b.  ➡ ①

3. Kim has 5 cents in her hand. Cross off (x) the hand that cannot be Kim's.

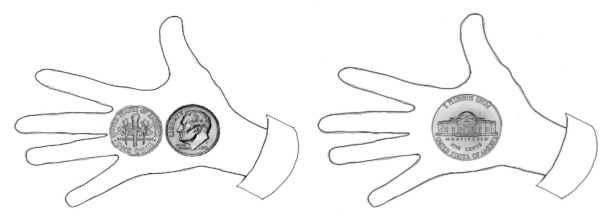

254   Lesson 20:   Identify pennies, nickels, and dimes by their image, name, or value. Decompose the values of nickels and dimes using pennies and nickels.

4. Anton has 10 cents in his pocket. One of his coins is a nickel. Draw coins to show two different ways he could have ten cents with the coins he has in his pocket.

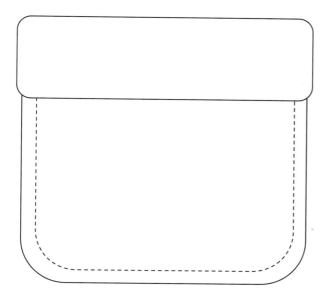
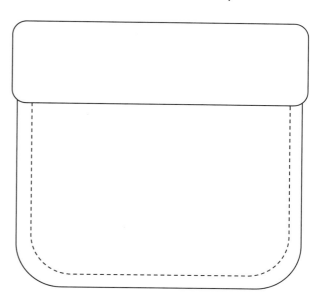

5. Emi says she has more money than Kiana. Is she correct? Why or why not?

Emi's Money

Kiana's Money

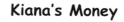

Emi is correct/not correct because _____

_____

A STORY OF UNITS   **Lesson 20 Exit Ticket**   **1•6**

Name _____ Date _____

1. Match the pennies to the coin with the same value.

   a.

   b.

2. Ben has 10 cents. He has 1 nickel. Draw more coin(s) to show what other coin(s) he might have.

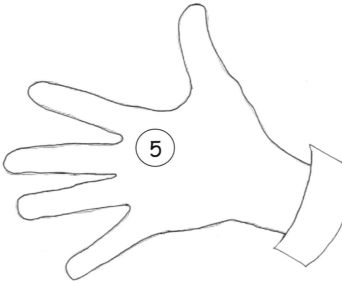

A STORY OF UNITS

Lesson 20 Homework 1•6

Name _____ Date _____

1. Match.

   •   penny   •

   •   nickel   •

   •   dime   •

2. Cross off some pennies so the remaining pennies show the value of the coin to their left.

   a.

   b.

Lesson 20: Identify pennies, nickels, and dimes by their image, name, or value. Decompose the values of nickels and dimes using pennies and nickels.

257

3. Maria has 5 cents in her pocket. Draw coins to show two different ways she could have 5 cents.

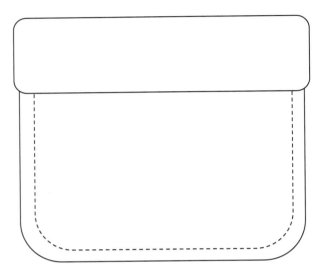
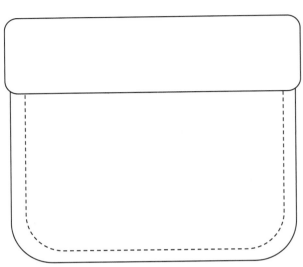

4. Solve. Draw a line to match the number sentence with the coin (or coins) that give the answer.

a. 10 cents + 10 cents = _____ cents   •   •

b. 10 cents - 5 cents = _____ cents   •   •

c. 20 cents - 10 cents = _____ cents   •   •

d. 9 cents - 8 cents = _____ cents   •   •

A STORY OF UNITS

Lesson 20 Template 1•6

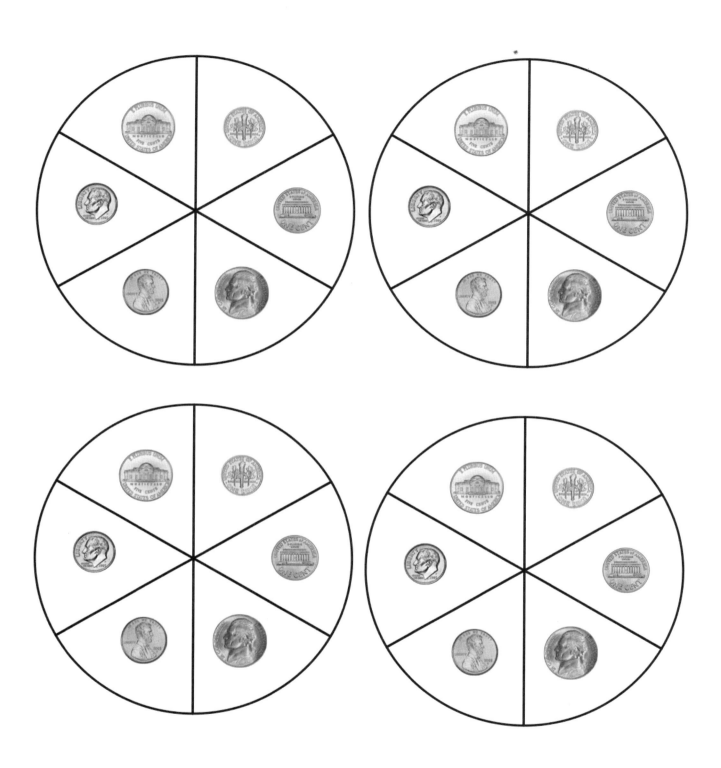

spinner: each group or set of partners needs 1 circle from this page. see image for use with pencil and paper clip.

Lesson 20: Identify pennies, nickels, and dimes by their image, name, or value. Decompose the values of nickels and dimes using pennies and nickels.

259

A STORY OF UNITS   Lesson 21 1•6

Lesson 21

Objective: Identify quarters by their image, name, or value. Decompose the value of a quarter using pennies, nickels, and dimes.

Suggested Lesson Structure

- ■ Fluency Practice   (10 minutes)
- ■ Application Problem   (5 minutes)
- □ Concept Development   (35 minutes)
- ■ Student Debrief   (10 minutes)
   Total Time   **(60 minutes)**

Fluency Practice (10 minutes)

- Grade 1 Core Fluency Sprint **1.OA.6**   (10 minutes)

Grade 1 Core Fluency Sprint (10 minutes)

Materials: (S) Core Fluency Sprints (Lesson 3)

Note: Based on the needs of the class, select a Sprint. There are several possible options available.

1. Re-administer the Sprint from the previous day's lesson.
2. Administer the next Sprint in the sequence.
3. Differentiate. Administer two different Sprints. Simply have one group do a counting activity on the back of the Sprint while the other Sprint is corrected.

Application Problem (5 minutes)

Willie saw 11 monkeys at the zoo. He saw 4 fewer monkeys than tigers. How many tigers did he see at the zoo?

Note: Today's problem is a *compare with bigger unknown* where the problem suggests the wrong operation. Students are expected to have worked with these problems in Grade 1, but mastery is not expected until the end of Grade 2. Consider scaffolding such as, "Set up your tape diagram to first show the same number of monkeys and tigers. Which animal did Willie see more of, monkeys or tigers? Add another section of tape (the *more* tape) to the tigers. How many more tigers than monkeys did Willie see?"

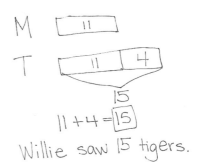

A STORY OF UNITS   Lesson 21   1•6

Concept Development (35 minutes)

Materials: (T) 4 quarters, 5 dimes, 5 nickels, 25 pennies (plastic or real), chart paper (S) 1 quarter, 3 dimes, 3 nickels, 25 pennies (plastic or real), 1 die per set of partners, Problem Set

Note: Prepare the chart shown to the right prior to the lesson.

Gather students in the meeting area with their coins. Hold the Problem Set to the side.

T: Sort your coins into piles like we did yesterday so that we can find the coins we want more quickly. (Wait as students sort their coins.)

T: Put your finger on one dime. What is the value of one dime?

S: 10 cents!

T: Put your finger on one penny. What is the value of one penny?

S: 1 cent!

T: Put your finger on one nickel. What is the value of one nickel?

S: 5 cents!

T: What is the unit for each of these coins?

S: Cents!

T: You have 1 new coin. Pick up the new coin. Look at it closely, and describe what you notice about this coin.

S: It's bigger than the other coins. → It has bumpy edges, like the dime. The penny and the nickel have smooth edges. → There is an eagle on this one. → This one has a state's name on it!

T: This coin is called a **quarter**. Let's all say *quarter*.

S: Quarter!

T: Some quarters have different images on the back. Many have eagles on them, but others have different pictures and names of the states on them. (Show a few different images of quarters.) But no matter what, a quarter has a value of 25 cents.

T: Let's use our coins to make 25 cents in different ways and record them on our chart.

T: How many pennies make 25 cents?

S: 25 pennies!

T: Count out 25 pennies. Please arrange them in 5-groups. I'll give you about one minute.

T: To draw 1 penny, we make a circle and write the value of the coin on it. (Demonstrate.) What is the value of 1 penny?

S: 1 cent.

T: Here is your chart. (Distribute the Problem Set to students.)

T: Quickly draw one penny, and show me your work. (Check students' work.) Now you have about one minute to draw 25 pennies in the first row of the Problem Set. Use the 5-group way.

T: How many tens do you see?

S: 2.

Lesson 21:   Identify quarters by their image, name, or value. Decompose the value of a quarter using pennies, nickels, and dimes.

A STORY OF UNITS   **Lesson 21 1•6**

T: How many ones do you see?
S: 25.
T: How many ones are not grouped in a ten?
S: 5.
T: Go down one row. What coins do we want to use to make 25 cents now?
S: Dimes and pennies!
T: Look at your 25 pennies without touching them. What is a way to trade to make 25 cents with dimes and pennies? Talk to your partner.
S: I can trade 10 pennies for 1 dime. → I can trade 20 pennies for 2 dimes. → I could put 2 dimes and 5 pennies. → I can put 1 dime and 15 pennies.
T: Go ahead and change pennies for dimes. Put the dimes where the pennies used to be. (Allow time for students to work.)

MP.4

T: To draw 1 dime, we make a circle and write the value of the dime on it. What is the value of 1 dime?
S: 10 cents.
T: What will you draw on the circle to show a dime?
S: 10.
T: Record one way you used dimes and pennies to make 25 cents.
S: (Record.)
T: Which was simpler, drawing 25 pennies or the dimes and pennies?
S: The dimes and pennies!
T: If you are ready to do the rest of the problems on your own in the chart, you may return to your desk with your coins and Problem Set. I will continue working here on the carpet with those who want to work together.

NOTES ON MULTIPLE MEANS OF ENGAGEMENT:

To immerse students in coins more fully, consider a classroom economy program for the duration of the year. Provide students with plastic or real coins for completing their classroom tasks. The money earned can be pooled toward a class goal or used individually in a class store.

Continue the process, emphasizing systematic trading and inviting alternate ways to use the coins indicated. Close by returning to the quarter.

T: How many quarters make 25 cents?
S: 1.
T: (Write 1 before *quarter,* draw a circle, and write 25 within it on the last row of the Problem Set.)
T: What is the easiest coin to use to show 25 cents?
S: A quarter!
T: Take a moment to review with your partner all the ways that you showed that have the same value as a quarter.

NOTES ON MULTIPLE MEANS OF EXPRESSION:

For students who need visual reminders of the names and values of the coins, hang chart paper with the name, value, and image of each coin.

Optional Activity: Engage students in a game of 25 Cents. The object of the game is to be the first player to exchange their money for 1 quarter.

---

262   Lesson 21:   Identify quarters by their image, name, or value. Decompose the value of a quarter using pennies, nickels, and dimes.

©2015 Great Minds. eureka-math.org
G1-M6-TE-BK6-1.3.1-1.2016

A STORY OF UNITS

Lesson 21 1•6

Materials: One die; 25 pennies, nickels, dimes, and quarters for trading; and a pot per pair of students

- Put all coins in a pot between the partners.
- Player A rolls the die and takes that number of pennies.
- Player B rolls the die and does the same.
- On each turn, players roll the die, add the additional pennies, and exchange their pennies for larger coins, if possible. For instance, if Player A has 6 pennies, she may trade 5 pennies for 1 nickel. If Player B has 1 nickel and 5 pennies, he may trade the coins for 1 dime.
- Play continues until a player can exchange his coins for 1 quarter, explaining that he has 25 cents.

Problem Set (10 minutes)

Students should do their personal best to complete the Problem Set within the allotted 10 minutes. For some classes, it may be appropriate to modify the assignment by specifying which problems they work on first. Some problems do not specify a method for solving. Students should solve these problems using the RDW approach used for Application Problems.

Student Debrief (10 minutes)

Lesson Objective: Identify quarters by their image, name, or value. Decompose the value of a quarter using pennies, nickels, and dimes.

The Student Debrief is intended to invite reflection and active processing of the total lesson experience.

Invite students to review their solutions for the Problem Set. They should check work by comparing answers with a partner before going over answers as a class. Look for misconceptions or misunderstandings that can be addressed in the Debrief. Guide students in a conversation to debrief the Problem Set and process the lesson.

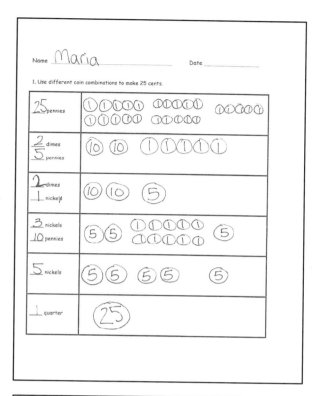

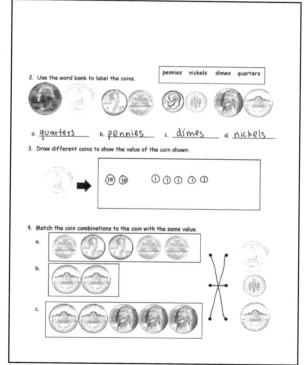

Lesson 21: Identify quarters by their image, name, or value. Decompose the value of a quarter using pennies, nickels, and dimes.

263

Any combination of the questions below may be used to lead the discussion.

- Look at Problem 4. How many more nickels did you need to make 25 cents than you needed to make 10 cents?
- What attributes of the coins help you recognize each?
- What is the name of the coin that has a value of 25 cents? (**Quarter**.)
- Where do you see quarters? What coins could you use to buy a snack that costs 55 cents?

Exit Ticket (3 minutes)

After the Student Debrief, instruct students to complete the Exit Ticket. A review of their work will help with assessing students' understanding of the concepts that were presented in today's lesson and planning more effectively for future lessons. The questions may be read aloud to the students.

A STORY OF UNITS

Lesson 21 Problem Set 1•6

Name _____ Date _____

1. Use different coin combinations to make 25 cents.

a.	____ pennies	
b.	____ dimes ____ pennies	
c.	____ dimes ____ nickels	
d.	____ nickels ____ pennies	
e.	____ nickels	
f.	____ quarter	

Lesson 21: Identify quarters by their image, name, or value. Decompose the value of a quarter using pennies, nickels, and dimes.

2. Use the word bank to label the coins. | pennies   nickels   dimes   quarters |

a. _____   b. _____   c. _____   d. _____

3. Draw different coins to show the value of the coin shown.

4. Match the coin combinations to the coin with the same value.

a.

b.

c.

A STORY OF UNITS

Lesson 21 Exit Ticket 1•6

Name _____   Date _____

Use the word bank to write the names of the coins.

dimes   nickels   pennies   quarters

a. _____   b. _____   c. _____   d. _____

A STORY OF UNITS

Lesson 21 Homework 1•6

Name _____ Date _____

1. Use the word bank to label the coins.

| dimes   nickels   pennies   quarters |

a. _____ b. _____ c. _____ d. _____

2. Write the value of each coin.

   a. The value of one dime is _____ cent(s).

   b. The value of one penny is _____ cent(s).

   c. The value of one nickel is _____ cent(s).

   d. The value of one quarter is _____ cent(s).

3. Your mom said she will give you 1 nickel or 1 quarter. Which would you take, and why?

_____

_____

_____

_____

Lesson 21: Identify quarters by their image, name, or value. Decompose the value of a quarter using pennies, nickels, and dimes.

4. Lee has 25 cents in his piggy bank. Which coin or coins could be in his bank?

   a. Draw to show the coins that could be in Lee's bank.

   b. Draw a different set of coins that could be in Lee's bank.

Lesson 21: Identify quarters by their image, name, or value. Decompose the value of a quarter using pennies, nickels, and dimes.

| A STORY OF UNITS | Lesson 22 1•6 |

Lesson 22

Objective: Identify varied coins by their image, name, or value. Add one cent to the value of any coin.

Suggested Lesson Structure

- ■ Fluency Practice   (13 minutes)
- ■ Application Problem (5 minutes)
- ■ Concept Development (32 minutes)
- ■ Student Debrief   (10 minutes)
-   **Total Time**   **(60 minutes)**

Fluency Practice (13 minutes)

- Core Fluency Differentiated Practice Sets **1.OA.6**   (5 minutes)
- Standards Check: Addition Within 20 **1.OA.6**   (8 minutes)

Core Fluency Differentiated Practice Sets (5 minutes)

Materials: (S) Core Fluency Practice Sets (Lesson 1)

Note: Give the appropriate Practice Set to each student. Students who completed all questions correctly on their most recent Practice Set should be given the next level of difficulty. All other students should try to improve their scores on their current levels.

Students complete as many problems as they can in 90 seconds. Assign a counting pattern and start number for early finishers, or have them practice make ten addition or subtraction on the back of their papers. Collect and correct any Practice Sets completed within the allotted time.

Standards Check: Addition Within 20 (8 minutes)

Materials: (S) Personal white board

Note: This fluency activity shows which strategies students are using to add within 20. Students may show their work with a number bond, the arrow way, multi-step equations, or listing numbers to show how to count on.

Write the following list of strategies:

1. Count all.
2. Count on.
3. Make ten.

A STORY OF UNITS

Lesson 22 1•6

4. Use a doubles fact.
5. Use a helper problem (e.g., to solve 15 + 3, add 5 and 3 first).

Say an addition expression. Students use their personal boards to solve. Choose students who used different strategies to share what they did, or instruct students to share their strategies with a partner.

Suggested sequence:

- 9 + 2, 3 + 9, 2 + 8 + 2
- 5 + 6, 7 + 6, 4 + 4 + 6
- 15 + 1, 3 + 16
- 13 + 4, 12 + 7

Application Problem (5 minutes)

Peter has 6 more red pencils than blue pencils. He has 8 blue pencils. How many red pencils does he have?

Note: Today's problem is a *compare with bigger unknown* problem type. Because yesterday's Application Problem suggested an incorrect operation, students may expect the same experience with today's problem. Encourage students to read through the entire problem, checking that their drawings and solutions make sense for all sentences in the story problem. Having students check their work helps them to become better problem solvers. Be sure to point this out.

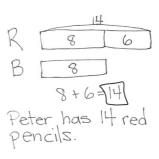

Concept Development (32 minutes)

Materials: (T) 5–10 different quarters (e.g., various commemorative quarters), 5 dimes, 5 nickels (possibly with different images), 20 pennies, 1 dollar coin if available (real or plastic), projector
(S) 1 quarter, 2–5 dimes, 3–5 nickels, 10–20 pennies (real or plastic), 1 die, coin spinner with quarter (Template), paper clip, pencil per pair, personal white board

Gather students in the meeting area with their materials. Distribute 1 or 2 coins to each student as they come to the meeting area.

T: I had all of these coins at home. Tell your partner the name and value of the coin(s) you have. Explain how you know what coin it is. (Wait as students share. Consider having them pass their coin to the right until each student has had a chance to identify all the major coins.)

T: Let's sort them into piles of the same coin. (Call out each coin. Students holding that type of coin place their coins in a common pile in the middle of the group.)

> **NOTES ON MULTIPLE MEANS OF ENGAGEMENT:**
>
> If a classroom economy has started, use students' coins to have them identify the image, name, and value of coins. Allow students to trade their pennies (or nickels) for different coins if they have enough to do so.

Lesson 22: Identify varied coins by their image, name, or value. Add one cent to the value of any coin.

T: (Point to the pennies.) What kind of coins are these?
S: Pennies!
T: What is the value of 1 penny?
S: 1 cent!
T: (Push forward 1 nickel.) What is the name of this coin?
S: It's a nickel.
T: What is its value?
S: 5 cents.
T: Use a complete sentence. A nickel's value is…?
S: A nickel's value is 5 cents.
T: (Push 1 penny next to the nickel.) If I have 1 nickel *and* 1 penny, how many cents do I have altogether?
S: 6 cents!
T: How do you know?
S: 5 + 1 = 6. → 5 cents plus one more cent is 6 cents.
T: (Draw 1 nickel and 1 penny on the chart paper, including their individual values and their total value, as shown to the right.)

MP.4

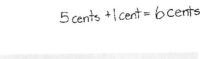

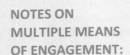

Repeat the process, first with 1 dime and 1 penny, and then with 1 quarter and 1 penny. Finally, push forward the dollar coin.

T: (Push forward a 1 dollar coin.) Does anyone know the name of this coin?
S: It's a **dollar coin**! (If students do not know, introduce this as a dollar coin.)
T: A dollar coin is worth 100 cents!
T: (Push 1 penny next to the dollar coin.) If I have a 1 dollar coin whose value is 100 cents and 1 penny, how many cents do I have altogether?
S: 101 cents!
T: (Add the dollar coin and penny to the chart paper, including their individual values and their total value.)

NOTES ON MULTIPLE MEANS OF ENGAGEMENT:

Engage students in a hunt at home for quarters, pennies, and nickels with various images. When students bring in their findings, have them sort and name each coin and its value. Encourage students to share interesting observations.

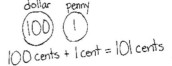

Based on students' ability to identify the name, value, and image of each coin, choose one of the games played during the past two days. To practice coin values of the dime, nickel, and penny, play Coin Trade. To practice adding on coins as well as trading coins, play 25 Cents.

Coin Trade

If students are ready, include the quarter, and use the new spinner at the end of this lesson.

The object of the game is to continue to trade coins, always having 10 cents.

Materials: Each player has 10 pennies (25 pennies, if using the new spinner) and the spinner with a paper clip and pencil; each pair has a pot with pennies, nickels, and dimes (and quarters if using the new spinner) for trading per pair.

- Partner A spins the spinner.
- Partner A trades pennies for the coin landed on. (For instance, if the student lands on a nickel, he trades 5 pennies for 1 nickel. If he lands on a dime, he trades 10 pennies for 1 dime. If he lands on a penny, he trades a penny for a penny.) Player A counts his coins to be sure he still has 10 cents.
- Partner B takes a turn. Player B counts her coins to be sure she still has 10 cents.
- Play continues as time allows.
- The person with the most pennies at the end of the game is the winner.

As play continues, students might land on the coins they already have, such as landing on a penny when they have 10 pennies. Students may trade one of their pennies for a new penny. Play the game for about five minutes or as time allows.

25 Cents

The object of the game is to be the first player to exchange their money for 1 quarter. For students who are ready for greater challenges, you can choose to make the goal 50 cents or 100 cents.

Materials: One die; 25 pennies, 5 nickels, 3 dimes and 2 quarters for trading; and a pot per pair of students

- Put all coins in a pot in the middle.
- Player A rolls the die and takes that number of pennies.
- Player B rolls the die and does the same.
- On each turn, players roll the die, add the additional pennies, and exchange their pennies for larger coins, if possible. For instance, if Player A has 6 pennies, he may trade 5 pennies for 1 nickel. If Player B has 1 nickel and 5 pennies, she may trade the coins for 1 dime.
- Play continues until a player can exchange his coins for 1 quarter, explaining that he has 25 cents.

Problem Set (10 minutes)

Students should do their personal best to complete the Problem Set within the allotted 10 minutes. For some classes, it may be appropriate to modify the assignment by specifying which problems they work on first. Some problems do not specify a method for solving. Students should solve these problems using the RDW approach used for Application Problems.

Lesson 22: Identify varied coins by their image, name, or value. Add one cent to the value of any coin.

A STORY OF UNITS   Lesson 22   1•6

Student Debrief (10 minutes)

Lesson Objective: Identify varied coins by their image, name, or value. Add one cent to the value of any coin.

The Student Debrief is intended to invite reflection and active processing of the total lesson experience.

Invite students to review their solutions for the Problem Set. They should check work by comparing answers with a partner before going over answers as a class. Look for misconceptions or misunderstandings that can be addressed in the Debrief. Guide students in a conversation to debrief the Problem Set and process the lesson.

Any combination of the questions below may be used to lead the discussion.

- Look at Problem 2. What other combinations of coins could you use to have the same value as a quarter? As a dime? As a nickel?
- Look at Problem 3. What are some ways to tell a nickel from a quarter?
- Create other problems like those in Problem 5. Who can identify the coin with the same value?
- What new coin did we see today? (**Dollar coin.** If applicable) Have you seen the dollar coin before? Where have you seen or used it?

Exit Ticket (3 minutes)

After the Student Debrief, instruct students to complete the Exit Ticket. A review of their work will help with assessing students' understanding of the concepts that were presented in today's lesson and planning more effectively for future lessons. The questions may be read aloud to the students.

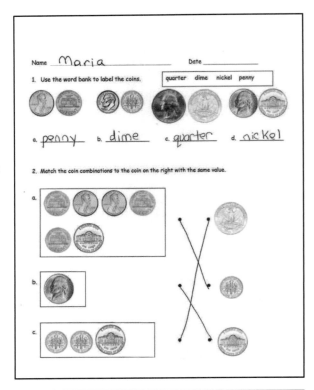

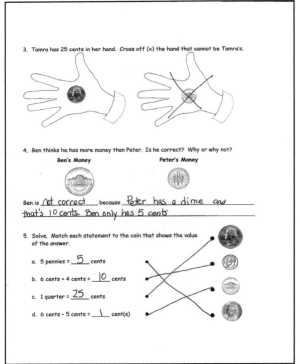

274   Lesson 22:   Identify varied coins by their image, name, or value. Add one cent to the value of any coin.

Name _____ Date _____

1. Use the word bank to label the coins. | quarter   dime   nickel   penny |

a. _____   b. _____   c. _____   d. _____

2. Match the coin combinations to the coin on the right with the same value.

a.    •   •

b.    •   •

c.    •   •

A STORY OF UNITS　　　　　　　　　　　　　　　　　　　　　　　Lesson 22 Problem Set 1•6

3. Tamra has 25 cents in her hand. Cross off (x) the hand that cannot be Tamra's.

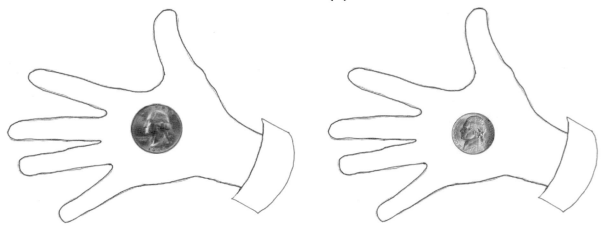

4. Ben thinks he has more money than Peter. Is he correct? Why or why not?

　　　　　　　　Ben's Money　　　　　　　　　　　Peter's Money

Ben is _____ because _____

_____

5. Solve. Match each statement to the coin that shows the value of the answer.

a. 5 pennies = _____ cents　　　　•　　　•　

b. 6 cents + 4 cents = _____ cents　•　　•　

c. 1 quarter = _____ cents　　　　•　　　•

d. 6 cents - 5 cents = _____ cent(s)　•　•　

Lesson 22: Identify varied coins by their image, name, or value. Add one cent to the value of any coin.

Name _____   Date _____

Draw a line to match each coin to its correct name.

   • | dime | •

   • | nickel | •

   • | penny | •

   • | quarter | •

A STORY OF UNITS

Lesson 22 Homework 1•6

Name _____ Date _____

1. Match the label to the correct coins, and write the value. There will be more than one match for each coin name.

   a. nickel

   _____ cents

   b. dime

   _____ cents

   c. quarter

   _____ cents

   d. penny

   _____ cent

Lesson 22: Identify varied coins by their image, name, or value. Add one cent to the value of any coin.

2. Lee has one coin in his pocket, and Pedro has 3 coins. Pedro has more money than Lee. Draw a picture to show the coins each boy might have.

Lee's Pocket

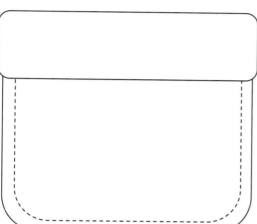

Pedro's Pocket

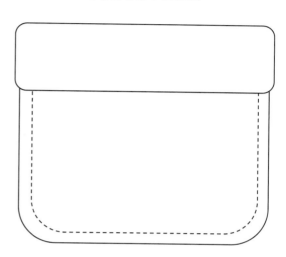

3. Bailey has 4 coins in her pocket, and Ingrid has 4 coins. Ingrid has more money than Bailey. Draw a picture to show the coins each girl might have.

Bailey's Pocket

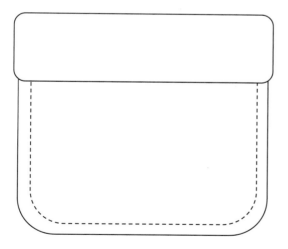

Ingrid's Pocket

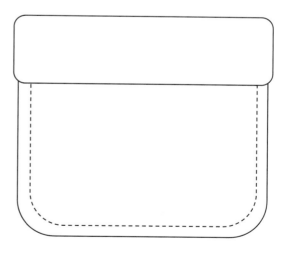

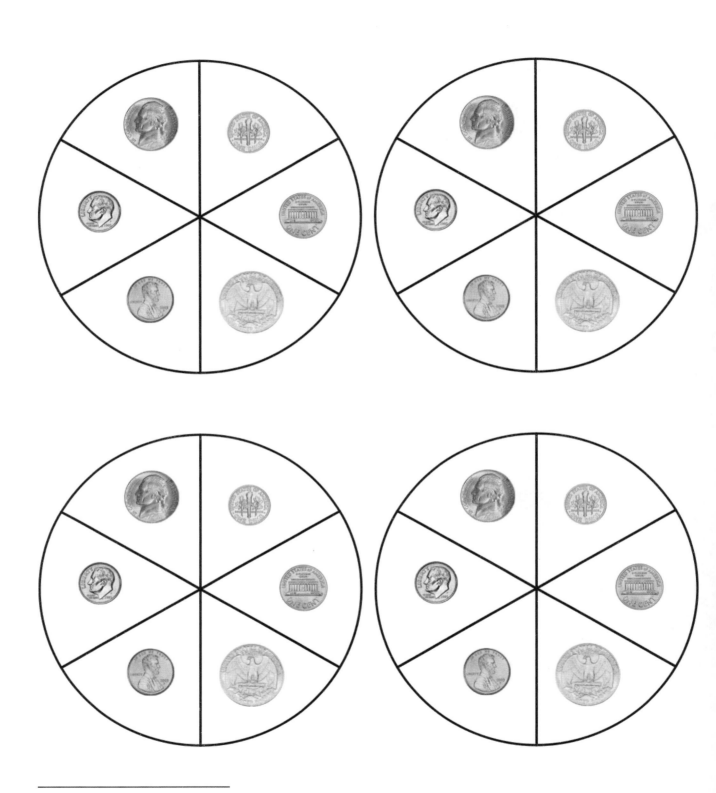

coin spinner with quarter

Lesson 23

Objective: Count on using pennies from any single coin.

Suggested Lesson Structure

- **Fluency Practice** (13 minutes)
- **Application Problem** (5 minutes)
- **Concept Development** (32 minutes)
- **Student Debrief** (10 minutes)
- **Total Time** **(60 minutes)**

Fluency Practice (13 minutes)

- Core Fluency Differentiated Practice Sets **1.OA.6**   (5 minutes)
- Standards Check: Subtraction Within 20 **1.OA.6**   (8 minutes)

Core Fluency Differentiated Practice Sets (5 minutes)

Materials: (S) Core Fluency Practice Sets (Lesson 1)

Note: Give the appropriate Practice Set to each student. Students who completed all questions correctly on their most recent Practice Set should be given the next level of difficulty. All other students should try to improve their scores on their current levels.

Students complete as many problems as they can in 90 seconds. Assign a counting pattern and start number for early finishers, or have them practice make ten addition or subtraction on the back of their papers. Collect and correct any Practice Sets completed within the allotted time.

Standards Check: Subtraction Within 20 (8 minutes)

Materials: (S) Personal white board

Note: This fluency activity shows which strategies students are using to subtract within 20. Students may show their work with a number bond, the arrow way, multi-step equations, or listing numbers to show how to count on.

Write the following list of strategies:

1. Count on or back.
2. Think of the addition problem.
3. Take from ten.
4. Use place value and a helper problem.

A STORY OF UNITS   Lesson 23 1•6

Say a subtraction expression. Students use their personal white boards to solve. Choose students who used different strategies to share what they did, or instruct students to share their strategies with a partner.

Suggested sequence:
- 15 – 1, 18 – 2
- 18 – 4, 19 – 7
- 12 – 3, 11 – 2
- 15 – 9, 17 – 8
- 16 – 14, 18 – 15

Application Problem (5 minutes)

Peter has 8 more green crayons than yellow crayons. Peter has 10 green crayons. How many yellow crayons does Peter have?

Note: Today's problem is a *compare with smaller unknown* where the problem suggests the wrong operation. Students are expected to have worked with these problems in Grade 1, but mastery is not expected until the end of Grade 2. Consider scaffolding such as, "Set up your tape diagram to first show the same number of green crayons and yellow crayons. Does Peter have more green crayons or yellow crayons? Add another section of tape (the *more* tape) to the green crayons. How many more green crayons does he have than yellow crayons?"

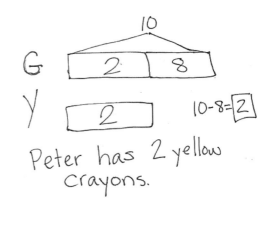

Concept Development (32 minutes)

Materials: (T) 1 quarter, 3–5 dimes, 2–4 nickels, 15 pennies (plastic or real), projector (S) 1 quarter, 3–5 dimes, 2–5 nickels, 25 pennies (plastic or real), 1 die per pair of students

Gather students in the meeting area with personal boards. Coins and dice are not needed until students play the game toward the end of the Concept Development.

T: (Project 1 quarter.) What is the name of this coin?
S: A quarter!
T: What is its value?
S: 25 cents.
T: (Add 1 penny to the quarter being projected.) How much money is shown now?
S: 26 cents!
T: How do you know?
S: You added one penny. That's one cent more.
T: What is 1 quarter plus 1 penny, a quarpenny? No such thing! But we can add their *values*! Let's try.

Lesson 23: Count on using pennies from any single coin.

A STORY OF UNITS

Lesson 23 1•6

T: Tell me an addition sentence that puts together the value of the quarter and the value of the penny.
S: 25 + 1 = 26.
T: Tell me an addition sentence that puts together the value of a dime and the value of 3 pennies.
S: 10 + 3 = 13.
T: So, a dime and 3 pennies would be how much money?
S: 13 cents.
T: Try some more!

Repeat the process by projecting the following sequence of coins:

- 1 quarter, add 3 pennies
- 3 dimes, add 6 pennies
  (Use 5-group formation to show the 6 pennies. Discuss why the 5-group formation helps students know the total amount of pennies without counting.)
- 1 nickel, add 4 pennies
- 4 pennies, add 1 nickel
  (Have students explain which coins they counted first and why. Accept both preferences.)

Practice counting on pennies using the following sequence:

- 3 pennies, 1 nickel
- 3 pennies, 1 quarter
- 4 pennies, 1 quarter

T: (Show 1 penny, 1 dime, 4 pennies.) How can we group these to make it easier to count?
S: Put all the pennies together!
T: Great! Which will we be starting with, the dime or the pennies?
S: The dime!
T: That is just easier; I agree. So, let's move all the pennies together and place them after the dime. (Move the first penny next to the 4 pennies.)
T: Tell me an addition sentence that puts together the value of a dime, the value of 4 pennies, and the value of 1 penny.
S: 10 + 4 + 1 = 15.

> **NOTES ON MULTIPLE MEANS OF EXPRESSION:**
>
> Some students may have difficulty keeping track of counted and uncounted coins. Invite students to place their own coins out to match the teacher's set of coins. Using these coins, students may rearrange the coins or slide the coins over as they count.

> **NOTES ON MULTIPLE MEANS OF ENGAGEMENT:**
>
> Have students, who may have difficulty keeping track of their total coin values between turns, use their boards to keep track of their totals as they play.

Continue to practice counting on pennies, regardless of the order of the coins using the following sequence:

- 2 pennies, 1 dime, 2 pennies
- 2 pennies, 1 quarter, 3 pennies
- 1 quarter, 7 pennies
  (Be sure to use the 5-group formation when presenting the 7 pennies. Discuss how the formation can help students use the make ten strategy to add.)

Lesson 23: Count on using pennies from any single coin.

283

A STORY OF UNITS   Lesson 23 1•6

Note: If time permits, have partners play First to 50 Cents (a version of Coin Exchange). The objective of the game is to be the first player with 50 cents.

First to 50 Cents

Players A and B each begin with 1 quarter.

1. Player A rolls the die and adds that many pennies to his quarter.
2. Player B rolls the die and adds that many pennies to her quarter.
3. Players continue to take turns until someone has at least 50 cents, trading pennies for nickels or dimes. No player who has 25 pennies can win!

Players might trade pennies for nickels, dimes, and finally a quarter as they play.

Problem Set (10 minutes)

Students should do their personal best to complete the Problem Set within the allotted 10 minutes. For some classes, it may be appropriate to modify the assignment by specifying which problems they work on first. Some problems do not specify a method for solving. Students should solve these problems using the RDW approach used for Application Problems.

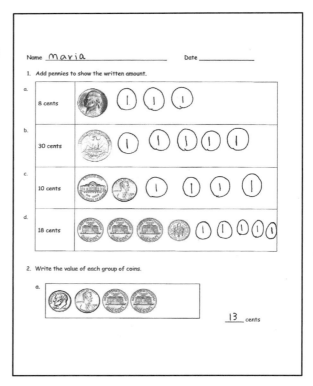

Student Debrief (10 minutes)

Lesson Objective: Count on using pennies from any single coin.

The Student Debrief is intended to invite reflection and active processing of the total lesson experience.

Invite students to review their solutions for the Problem Set. They should check work by comparing answers with a partner before going over answers as a class. Look for misconceptions or misunderstandings that can be addressed in the Debrief. Guide students in a conversation to debrief the Problem Set and process the lesson.

Any combination of the questions below may be used to lead the discussion.

- Look at Problem 2. How do 5-group formations help you count coins quickly?
- Three dimes and 1 dime is 4 dimes. Three pennies and 1 penny is 4 pennies. Why is it that 3 dimes and 1 penny don't equal 4 cents? What do we need to do in order to add dimes and pennies together? What is our label, or unit, to add 3 dimes and 1 penny in a number sentence?
 (30 cents + 1 cent = 31 cents. We change the unit to cents so that they have the same unit, which can be added together.)

284   Lesson 23:   Count on using pennies from any single coin.

- Look at Problem 2(b). How many cents are there? Look at Problem 2(c). How many cents are there? Why is the value of the coins in Problem 2(c) greater than the value of the coins in Problem 2(b) even though there are more coins in Problem 2(b)?

Exit Ticket (3 minutes)

After the Student Debrief, instruct students to complete the Exit Ticket. A review of their work will help with assessing students' understanding of the concepts that were presented in today's lesson and planning more effectively for future lessons. The questions may be read aloud to the students.

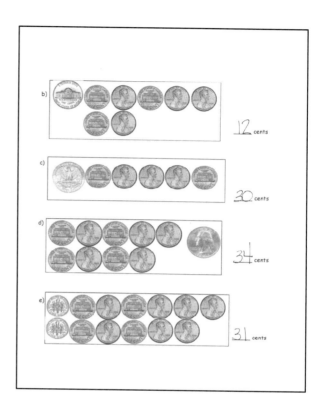

Lesson 23: Count on using pennies from any single coin.

Name _____ Date _____

1. Add pennies to show the written amount.

2. Write the value of each group of coins.

a.

_____ cents

Lesson 23 Problem Set 1•6

b.  _____ cents

c.  _____ cents

d.  _____ cents

e.  _____ cents

Lesson 23: Count on using pennies from any single coin.

A STORY OF UNITS

Lesson 23 Exit Ticket 1•6

Name _____ Date _____

Add pennies to show the written amount.

a.	9 cents	
b.	29 cents	

Name _____ Date _____

1. Add pennies to show the written amount.

a.	15 cents	(dime)
b.	28 cents	(quarter)
c.	22 cents	(dime) (nickel)
d.	32 cents	(dime) (nickel) (dime)

2. Write the value of each group of coins.

a.

_____ cents

Lesson 23: Count on using pennies from any single coin.

b.

_____ cents

c.

_____ cents

d.

_____ cents

e.

_____ cents

Lesson 23: Count on using pennies from any single coin.

A STORY OF UNITS   Lesson 24 1•6

Lesson 24

Objective: Use dimes and pennies as representations of numbers to 120.

Suggested Lesson Structure

- ■ Fluency Practice   (13 minutes)
- ■ Application Problem   (5 minutes)
- ■ Concept Development   (32 minutes)
- ■ Student Debrief   (10 minutes)
- **Total Time**   **(60 minutes)**

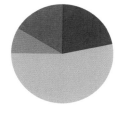

Fluency Practice (13 minutes)

- Grade 1 Core Fluency Sprint **1.OA.6**   (10 minutes)
- Standards Check: Place Value **1.NBT.2**   (3 minutes)

Grade 1 Core Fluency Sprint (10 minutes)

Materials: (S) Core Fluency Sprints (Lesson 3)

Note: Choose a Sprint based on the needs of the class.

- Core Addition Sprint 1
- Core Addition Sprint 2
- Core Subtraction Sprint
- Core Fluency Sprint: Totals of 5, 6, and 7
- Core Fluency Sprint: Totals of 8, 9, and 10

Standards Check: Place Value (3 minutes)

Materials: (T/S) Personal white board

Note: This activity monitors students' understanding of place value.

Write a number on a personal white board, but do not show students.

   T: My number has 1 ten and 3 ones. What's my number?
   S: 13.
   T: (Show the board.) What's the value of this 1? (Pause, and then snap.)
   S: 10.

A STORY OF UNITS   Lesson 24 1•6

T: What's the value of this 3? (Pause, and then snap.)
S: 3.

Repeat with the following suggested sequence: 22, 27, 66, 63, 36, 90, and 99. Alternate saying the number in the ones place first and saying the number in the tens place first. For the last minute, write a two-digit number, and ask students to write the value of one of the digits on their personal white boards.

T: (Show 53.) Write the value of the 5.
S: (Write 50.)

Application Problem (5 minutes)

There are 8 eggs in the carton. The carton can hold 12 eggs. How many more eggs will fit in the carton?

Note: Today's problem is a *put together with addend unknown* problem type where students are looking for a missing part. A single bar is effective, especially since the problem is talking about one carton of eggs that looks like a single tape.

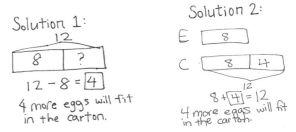

However, some students may want to model the problem with two bars, in a sense comparing the given part with the known total. This does make sense, especially considering they have been working with comparison problems recently. Both solutions are modeled.

Concept Development (32 minutes)

Materials: (T) 12 dimes, 10 pennies (plastic or real), projector (S) 12 dimes, 10 pennies (plastic or real), personal white board

Pair students. Pairs begin the lesson at their desks or tables with all materials.

T: (Write 80 on the board.) Use your coins to represent this number. Draw a matching place value chart on your personal white board.
S: (Use 8 dimes. Some may use 7 dimes and 10 pennies, which is considered correct as long as the student's place value chart matches his chosen representation.)
T: If I used only dimes to represent 80, how many dimes would I need?
S: 8 dimes!
T: How many tens are in 80?
S: 8 tens!

Repeat the process with the following suggested sequence: 50, 68, 82.

T: (Write 90 on the board.) Use your coins to represent this number. Draw a matching place value chart on your personal board.
S: (Use 9 dimes.)

292   Lesson 24:   Use dimes and pennies as representations of numbers to 120.

T:	If I used only dimes to represent 90, how many dimes would I need?	
S:	9 dimes!	
T:	How many tens are in 90?	
S:	9 tens!	
T:	(Write 92 on the board.) Use your coins to represent this number. Draw a matching place value chart on your personal white board.	
S:	(Use 9 dimes and 2 pennies.)	
T:	How many dimes would I need?	
S:	9 dimes!	
T:	How many pennies?	
S:	2 pennies!	
T:	How many tens and how many ones is this?	
S:	9 tens and 2 ones.	
T:	(Write 100 on the board.) How many tens are in 100? Use your dimes to show 100 cents. (Wait as students count out 10 dimes.)	
S:	(Show 10 dimes.)	
T:	How many dimes did we use to make 100 cents?	
S:	10 dimes!	
T:	How many tens do you have?	
S:	10 tens.	
T:	(Next to 100, add a place value chart showing 10 tens.)	
T:	Do we need any additional pennies?	
S:	No.	
T:	(Write 0 in the ones place on the place value chart.)	
T:	(Point to the place value chart.) 10 tens 0 ones is…?	
S:	100.	
T:	Let's add 1 more dime. (Wait as students add 1 dime to their collection.) How many dimes do you have now?	
S:	11 dimes!	
T:	Draw a place value chart on your personal white board to show 11 tens 0 ones. (Wait as students show this.)	
T:	(Write 100 + 10 on the board.) We added ten cents to one hundred cents. How many cents do we have now?	
S:	110 cents.	
T:	How many tens are in 110 cents?	
S:	11 tens!	
T:	Let's add 1 more dime. (Wait as students add 1 dime to their collection.) How many dimes do you have now?	
S:	12 dimes!	

Lesson 24: Use dimes and pennies as representations of numbers to 120.

| A STORY OF UNITS | Lesson 24 |

T: Draw a place value chart on your personal white board to show 12 tens 0 ones. (Wait as students show this.)

T: (Write 100 + 20 on the board.) We had 100 cents. Then, we added 2 more dimes for 20 more cents. How many cents do we have now?

S: 120 cents.

T: Look at your dimes. How many tens are in 120 cents?

S: 12 tens!

Note: Some students may be familiar with the value of a dollar and may bring up that 100 cents is 1 dollar or that 120 cents is $1.20. Let them know they are correct, but refocus them back to the number of tens (dimes) and ones (pennies), as that is the focus of this lesson.

Project the following sequences of coins, and have students determine their total value:

- 4 dimes, 8 pennies
- 4 dimes, 10 pennies
- 4 dimes, 12 pennies
- 5 pennies, 6 dimes
- 15 pennies, 6 dimes
- 10 dimes, 10 pennies

If students need more practice or support representing the numbers or the coins, continue presenting more two-digit numbers.

If students demonstrate strong skills in representing numbers to 120 using dimes and pennies, connect their understanding with their addition work from Topics C and D as shown below:

T: (Write 52 on the board.) Partner A, use your coins to represent this number using as many dimes as you can.

T: (Write 20 on the board.) Partner B, use your coins to represent this number using as many dimes as you can.

T: (Place an addition symbol between the numbers to create an expression.) Add your coins together. How much do you have? (Wait as students add the coins.)

S: 72 cents!

T: On your personal board, solve 52 + 20. (Wait as students solve.) How did you solve this problem?

S: I lined up my numbers and added the ones with ones and the tens with tens. There were only 2 ones. 5 tens + 2 tens is 7 tens. The total is 72. → I did the same thing. It's just like adding the dimes with the dimes. There were 2 pennies. Then, 5 dimes plus 2 dimes was 7 dimes. That makes 72 cents! → I added 2 tens. 52, 62, 72. → That's like counting on the dimes.

> **NOTES ON MULTIPLE MEANS OF REPRESENTATION:**
>
> Some students may have difficulty determining the value of coins when two different coins are used. Have them count one type of coin at a time and use their personal white boards to help them keep track of what they have counted.

> **NOTES ON MULTIPLE MEANS OF ENGAGEMENT:**
>
> Continue to challenge advanced students. As an extension to the lesson, add 2 or 4 nickels to the sequence to the left, and have students share their strategies to solve. They may count the nickels as nickels, count the nickels together as a ten, or ask to exchange two nickels for one dime.

Lesson 24: Use dimes and pennies as representations of numbers to 120.

Repeat the process using the following suggested sequence: 52 + 24, 59 + 30, 59 + 31, 59 + 34. As students share their solution strategies, ask them to make connections between their coins and their written notation. What similarities do they notice? What number bonds do they see represented by the coin combinations?

Problem Set (10 minutes)

Students should do their personal best to complete the Problem Set within the allotted 10 minutes. For some classes, it may be appropriate to modify the assignment by specifying which problems they work on first. Some problems do not specify a method for solving. Students should solve these problems using the RDW approach used for Application Problems.

Student Debrief (10 minutes)

Lesson Objective: Use dimes and pennies as representations of numbers to 120.

The Student Debrief is intended to invite reflection and active processing of the total lesson experience.

Invite students to review their solutions for the Problem Set. They should check work by comparing answers with a partner before going over answers as a class. Look for misconceptions or misunderstandings that can be addressed in the Debrief. Guide students in a conversation to debrief the Problem Set and process the lesson.

Any combination of the questions below may be used to lead the discussion.

- Look at Problem 2(a). How did you determine which set of 8 would make 80 cents? What is the value of the other set? How would a place value chart for 8 pennies look compared to the place value chart for 8 dimes?
- Look at Problem 2(b). What is the value of the set that does *not* equal 100 cents? How would you show this value in a place value chart?
- Look at Problem 3. What is another way to show 58 cents?

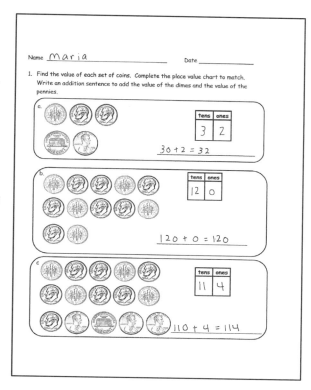

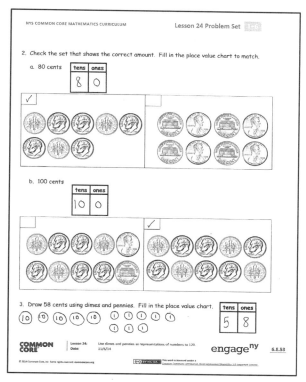

Lesson 24: Use dimes and pennies as representations of numbers to 120.

Exit Ticket (3 minutes)

After the Student Debrief, instruct students to complete the Exit Ticket. A review of their work will help with assessing students' understanding of the concepts that were presented in today's lesson and planning more effectively for future lessons. The questions may be read aloud to the students.

Name _____ Date _____

1. Find the value of each set of coins. Complete the place value chart to match. Write an addition sentence to add the value of the dimes and the value of the pennies.

a.

tens	ones

_____

b.

tens	ones

_____

c.

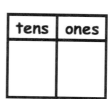

tens	ones

_____

Lesson 24: Use dimes and pennies as representations of numbers to 120.

2. Check the set that shows the correct amount. Fill in the place value chart to match.

   a. 80 cents

tens	ones

   b. 100 cents

tens	ones

3. Draw 58 cents using dimes and pennies. Fill in the place value chart.

tens	ones

Name _____   Date _____

Find the value of the set of coins. Complete the place value chart to match.
Write an addition sentence to add the value of the dimes and the value of the pennies.

tens	ones

_____

Lesson 24: Use dimes and pennies as representations of numbers to 120.

A STORY OF UNITS Lesson 24 Homework 1•6

Name _____ Date _____

1. Find the value of each set of coins. Complete the place value chart.
   Write an addition sentence to add the value of the dimes and the value of the pennies.

   a.

tens	ones

   _____

   b.

tens	ones

   _____

   c.

tens	ones

   _____

2. Check the set that shows the correct amount. Fill in the place value chart to match.

   110 cents

tens	ones

3. a. Draw 79 cents using dimes and pennies. Fill in the place value chart to match.

tens	ones

   b. Draw 118 cents using dimes and pennies. Fill in the place value chart to match.

tens	ones

A STORY OF UNITS

Mathematics Curriculum

1 GRADE

GRADE 1 • MODULE 6

Topic F
Varied Problem Types Within 20

1.OA.1

Focus Standard:	1.OA.1	Use addition and subtraction within 20 to solve word problems involving situations of adding to, taking from, putting together, taking apart, and comparing, with unknowns in all positions, e.g., by using objects, drawings, and equations with a symbol for the unknown number to represent the problem. (See CCSS-M Glossary, Table 1.)
Instructional Days:	3	
Coherence -Links from:	G1–M3	Ordering and Comparing Length Measurements as Numbers
	G1–M4	Place Value, Comparison, Addition and Subtraction to 40
-Links to:	G2–M7	Problem Solving with Length, Money, and Data

Topic F provides students the opportunity to focus on solving various problem types and to learn from their peers' strategies.

Lessons 25 and 26 focus on the most challenging Grade 1 problem types: *compare with bigger unknown* and *compare with smaller unknown* (**1.OA.1**). Students continue to strengthen their ability to recognize *compare* problem types and solve for unknowns in varied positions. They also work with problem types that suggest the incorrect operation, such as, "Shanika went down the slide 15 times. She went down 3 more times than Fran. How many times did Fran go down the slide?" While students do not need to master this problem type in Grade 1, exposure to these problems can support students' long-term success. During Lesson 26, students are provided more time to practice the various problem types and to learn to persevere in problem solving.

In Lesson 27, students practice all of the problem types they have encountered throughout the year. They discuss their methods for solving the problems and explain their work, including answering such questions as, "How does Student A's work help her solve the problem? How does Student B's work help him solve the problem? What compliment can we give Student A? What might Student A do to improve her work? What do you notice about your own work after looking at Student A's and Student B's work?"

Topic F

A Teaching Sequence Toward Mastery of Varied Problem Types Within 20

Objective 1: Solve *compare with bigger or smaller unknown* problem types.
(Lessons 25–26)

Objective 2: Share and critique peer strategies for solving problems of varied types.
(Lesson 27)

Lesson 25

Objective: Solve *compare with bigger or smaller unknown* problem types.

Suggested Lesson Structure

■ Fluency Practice   (15 minutes)
■ Concept Development (35 minutes)
■ Student Debrief   (10 minutes)
   Total Time   **(60 minutes)**

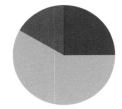

Fluency Practice (15 minutes)

- Grade 1 Core Fluency Sprint **1.OA.6**   (10 minutes)
- Standards Check: Add and Subtract Tens **1.NBT.4, 1.NBT.5**   (5 minutes)

Grade 1 Core Fluency Sprint (10 minutes)

Materials: (S) Core Fluency Sprint (Lesson 3)

Note: Based on the needs of the class, select a Sprint. There are several possible options available.

1. Re-administer the Sprint from the previous lesson.
2. Administer the next Sprint in the sequence.
3. Differentiate. Administer two different Sprints. Simply have one group do a counting activity on the back of the Sprint while the other Sprint is corrected.

Standards Check: Add and Subtract Tens (5 minutes)

Materials: (S) Personal white board

Note: This fluency activity monitors students' ability to add and subtract tens. All students must be able to find ten more or less than a number mentally.

   T: What's ten more than 25?
   S: 35.
   T: Write the number sentence.
   S: (Write 25 + 10 = 35.)
   T: What's ten less than 25?
   S: 15.

A STORY OF UNITS

Lesson 25 1•6

T: Write the number sentence.
S: (Write 25 – 10 = 15.)
T: Prove it. Draw quick tens and ones.
S: (Draw.)

Repeat with the suggested problem types. Alternate directing students to prove it with a number sentence, a number bond, or quick tens and ones. Include opportunities for students to prove a subtraction problem with an addition sentence (e.g., prove 10 less than 60 is 50 by writing 50 + 10 = 60).

- Mentally calculate 10 more/less than any two-digit number.
- Add and subtract multiples of 10 from multiples of 10 (e.g., 90 – 20; 40 + 50).
- Calculate multiples of 10 more than any two-digit number (e.g., 37 + 40).

Concept Development (35 minutes)

Materials: (T) Chart paper (S) Personal white board

Note: As students approach each problem, give them the opportunity to persevere and make sense of the problem on their own before intervening. When support is necessary, encourage the student to slow down and read each sentence carefully. During the Student Debrief, recognize students who have been successful at persevering.

Students sit in the meeting area or at their tables with their personal white boards.

Problems 1 and 2: *Compare with bigger or smaller unknown* problem types with *more* or *fewer* suggesting the correct operation.

T: Let's read our story together.
S/T: Ben played 9 songs on his banjo. Lee played 3 more songs than Ben. How many songs did Lee play?
T: On your personal white board, draw and then write a number sentence to match the story. (Circulate and observe students' solutions.)
S: (Draw and solve.)
T: (Choose a student who made a double tape diagram.) Tell us how you drew your tape diagram.
S: First, I made Ben's and Lee's tapes to be equal, but I know that's not true. Lee played 3 more songs. So I drew a *more* tape next to Lee's tape and wrote a 3 in it. Then, I put 9 in Ben's tape. I know Lee's first tape is 9 because it's the same size as Ben's tape. Lee's tape is now 9 and 3. That's 12 songs.
T: Excellent! What number sentence did we use to match the story?
S: 9 + 3 = 12.
T: What does the nine describe in the story and in our model? (Point.)
S: Ben's songs.

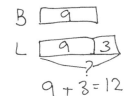

Lesson 25: Solve compare with bigger or smaller unknown problem types.

305

T: The three? (Point.)
S: The extra songs Lee played. → The 3 more songs of Lee.
T: The 12? (Point.)
S: How many songs Lee played.
T: Give me a statement answering the question.
S: Lee played 12 songs.

Repeat the process using the problem given below:

Nikil hopped on one leg 15 times in a row. Kim hopped 4 fewer times. How many times did Kim hop on one leg?

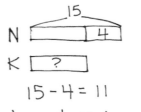

Problem 3: *Compare with smaller unknown* problem type with *more* suggesting the incorrect operation.

T: Let's read our next story problem together.
S/T: Shanika went down the slide 15 times. She went down 3 more times than Fran. How many times did Fran go down the slide?
T: Let's draw a double tape diagram since we need to find out how many times Fran went down the slide.
T: (Write S and F, and draw same-size tapes as shown to the right.) What do we need to ask ourselves first?
S: Who has more?
T: Yes! Read the story again carefully. (Wait.) Who has more? Who went down the slide more times?
S: Shanika!
T: (Draw a *more* tape next to Shanika's first tape.) How many more?
S: 3 more!
T: (Write 3 in the *more* tape.) Let's go back to the story and read the first sentence.
S/T: Shanika went down the slide 15 times.
T: Where should we put the 15? Turn and talk to your partner.
S: We can put it in the first part of Shanika's tape.
T: Who agrees? Who disagrees? (Choose a student who disagrees.) Tell us why. (Demonstrate as the student explains.)
S: If we put 15 in the first part of her tape, then it will show that Shanika went down the slide 18 times because her tape will show 15 and 3.
T: You're correct. That does not match the first sentence of our story problem, so where would we write 15?
S: Draw the arms to include both parts of Shanika's tape. The whole tape is 15.
T: (Demonstrate.) Yes! That makes sense! Let's read the second sentence.
S/T: She went down 3 more times than Fran.
T: Did we take care of that in our drawing? How?
S: Yes! We added a *more* tape for Shanika and wrote 3 inside.

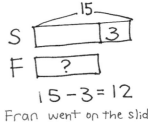

Lesson 25: Solve compare with bigger or smaller unknown problem types.

A STORY OF UNITS   Lesson 25 1•6

T: Let's read the last sentence.
S/T: How many times did Fran go down the slide?
T: Fran's tape gets the question mark since that's the unknown. Turn and talk to your partner about how you can solve Fran's amount.
S: We know that the first part of Shanika's tape is equal to Fran's tape, so we can just figure out Shanika's first part. → That's easy to do. We know the total is 15, and one part is 3. 15 – 3 gives us the other part. It's 12. → Shanika's first part is 12, so Fran's tape must be 12, too!
T: So, how many times did Fran go down the slide?
S: 12 times!
T: Take a moment to match the story to the model with your partner.
T: (Allow students sharing time.) What number sentence can we use to match this problem?
S: 15 – 3 = 12.
T: Tell your partner what each number in the sentence is telling about in the story, and then tell a statement that answers the question.
S: (Discuss referents.) Shanika went down the slide 12 times.

Repeat the process using the problem below:

Martha picked up 15 rocks on the beach. She picked up 8 more than Peter. How many rocks did Peter pick up at the beach?

Problem 4: *Compare with bigger unknown* problem type with *fewer* suggesting the incorrect operation.

T: Let's read the next story.
S/T: Anton caught 10 fireflies. He caught 7 fewer fireflies than Julio. How many fireflies did Julio catch?
T: Set up your tape diagram so it shows who the characters are. Make your tapes so they start out having the same amount.
S: (Draw two same-size tapes with labels A and J as shown to the right.)

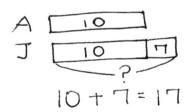

T: I love how you made each boy have equal-size tapes. But is this true?
S: No!
T: We have to ask…?
S: Who has more!
T: Okay! Read carefully and find out who has more. Then, add the *more* tape in your drawing.
S: (Develop the tape diagram as the teacher circulates and gives support.)
T: (Select a student who showed 7 more for Julio.)
S: I know that Julio has 7 more because the story said Anton caught 7 *fewer* fireflies, so I gave Julio the *more* tape and wrote 7 inside.

NOTES ON MULTIPLE MEANS OF ACTION AND EXPRESSION:

If students struggle with word problems, consider using either smaller numbers or encouraging students to include circle representations for the objects and then drawing rectangles around the circles to create the tape diagrams.

Lesson 25: Solve compare with bigger or smaller unknown problem types. 307

A STORY OF UNITS

Lesson 25 1•6

T: Excellent. Now that we have our tape diagram all set up, let's read the first sentence.
S/T: Anton caught 10 fireflies.
T: Decide where this information will go in your tape diagram.
S: (Write 10 in Anton's tape.)
T: Read the next sentence.
S: He caught 7 *fewer* fireflies than Julio.
T: Check your tape diagram. Did we include this information correctly?
S: Yes!
T: Explain to your partner how you showed this in your tape diagram.
S: Anton caught 7 fewer fireflies, so that means Julio caught 7 more. We added the *more* tape to Julio's first tape.
T: How many fireflies did Julio catch? Where does the question mark for the unknown go?
S: Under all of Julio's tape! → Draw arms to include both parts.
T: How many fireflies did Julio catch? Go ahead and solve. Turn and talk to your partner about how you got your answer.
S: (Solve and discuss.)
T: How did you find your answer?
S: I know that Julio's first part is the same as Anton's tape. That's 10. Julio had 7 more. So, 10 + 7 = 17. Julio caught 17 fireflies!
T: Excellent work. I'm especially proud of how carefully you read to find out who had more in every story.

NOTES ON MULTIPLE MEANS OF REPRESENTATION:
Some students may find it helpful to use linking cubes to represent the problems. Students can use different color linking cubes for each part being represented and then draw the tape diagrams to match their concrete representations.

Repeat the process using the following:

Darnel has 13 baseball cards. He has 4 fewer than Willie. How many baseball cards does Willie have?

Problem Set (10 minutes)

Students should do their personal best to complete the Problem Set within the allotted 10 minutes. For some classes, it may be appropriate to modify the assignment by specifying which problems they work on first. Some problems do not specify a method for solving. Students should solve these problems using the RDW approach used for Application Problems.

Student Debrief (10 minutes)

Lesson Objective: Solve *compare with bigger or smaller unknown* problem types.

The Student Debrief is intended to invite reflection and active processing of the total lesson experience.

Invite students to review their solutions for the Problem Set. They should check work by comparing answers with a partner before going over answers as a class. Look for misconceptions or misunderstandings that can be addressed in the Debrief. Guide students in a conversation to debrief the Problem Set and process the lesson.

Any combination of the questions below may be used to lead the discussion.

- How was setting up your tape diagram for Problem 4 different from Problem 5?
- Why is it easier to use a double tape diagram when we are comparing amounts?
- Why is it important to read every part of the story problem carefully? Give an example using your Problem Set or from today's lesson.
- Sometimes going slower when we do math means we are getting smarter. Find an example from your work today when you slowed down to get a problem correct.

Exit Ticket (3 minutes)

After the Student Debrief, instruct students to complete the Exit Ticket. A review of their work will help with assessing students' understanding of the concepts that were presented in today's lesson and planning more effectively for future lessons. The questions may be read aloud to the students.

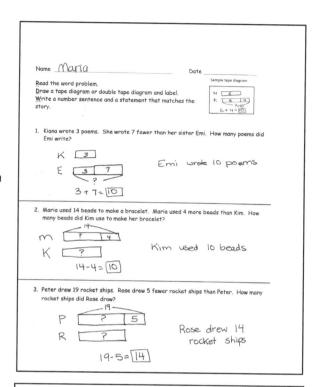

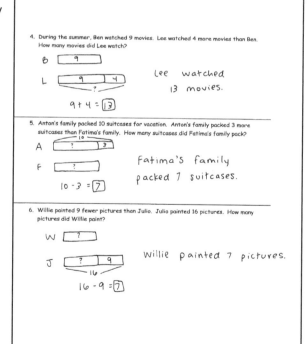

Lesson 25: Solve compare with bigger or smaller unknown problem types.

Name _____ Date _____

Read the word problem.
Draw a tape diagram or double tape diagram and label.
Write a number sentence and a statement that matches the story.

Sample Tape Diagram

1. Kiana wrote 3 poems. She wrote 7 fewer than her sister Emi. How many poems did Emi write?

2. Maria used 14 beads to make a bracelet. Maria used 4 more beads than Kim. How many beads did Kim use to make her bracelet?

3. Peter drew 19 rocket ships. Rose drew 5 fewer rocket ships than Peter. How many rocket ships did Rose draw?

4. During the summer, Ben watched 9 movies. Lee watched 4 more movies than Ben. How many movies did Lee watch?

5. Anton's family packed 10 suitcases for vacation. Anton's family packed 3 more suitcases than Fatima's family. How many suitcases did Fatima's family pack?

6. Willie painted 9 fewer pictures than Julio. Julio painted 16 pictures. How many pictures did Willie paint?

Name _____ Date _____

Sample Tape Diagram

Read the word problem.
Draw a tape diagram or double tape diagram and label.
Write a number sentence and a statement that matches the story.

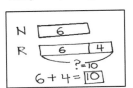

Willie splashed in 7 more puddles after the rainstorm than Julio. Willie splashed in 11 puddles. How many puddles did Julio splash in after the rainstorm?

A STORY OF UNITS

Lesson 25 Homework 1•6

Name _____ Date _____

Read the word problem.
Draw a tape diagram or double tape diagram and label.
Write a number sentence and a statement that matches the story.

Sample Tape Diagram

1. Julio listened to 7 songs on the radio. Lee listened to 3 more songs than Julio. How many songs did Lee listen to?

2. Shanika caught 14 ladybugs. She caught 4 more ladybugs than Willie. How many ladybugs did Willie catch?

3. Rose packed 3 more boxes than her sister to move to their new house. Her sister packed 11 boxes. How many boxes did Rose pack?

Lesson 25: Solve compare with bigger or smaller unknown problem types.

4. Tamra decorated 13 cookies. Tamra decorated 2 fewer cookies than Emi. How many cookies did Emi decorate?

5. Rose's brother hit 12 tennis balls. Rose hit 6 fewer tennis balls than her brother. How many tennis balls did Rose hit?

6. With his camera, Darnel took 5 more pictures than Kiana. He took 13 pictures. How many pictures did Kiana take?

| A STORY OF UNITS | Lesson 26 1•6 |

Lesson 26

Objective: Solve *compare with bigger or smaller unknown* problem types.

Suggested Lesson Structure

- ■ Fluency Practice   (15 minutes)
- ■ Concept Development (35 minutes)
- ■ Student Debrief   (10 minutes)
- **Total Time**   **(60 minutes)**

Fluency Practice (15 minutes)

- Core Fluency Differentiated Practice Sets **1.OA.6**   (5 minutes)
- Standards Check: Time **1.MD.3**   (5 minutes)
- Fluency Favorite or Standards Review   (5 minutes)

Core Fluency Differentiated Practice Sets (5 minutes)

Materials: (S) Core Fluency Practice Sets (Lesson 1)

Note: Give the appropriate Practice Set to each student. Students who completed all of the questions correctly on their most recent Practice Set should be given the next level of difficulty. All other students should try to improve their scores on their current levels.

Students complete as many problems as they can in 90 seconds. Assign a counting pattern and start number for early finishers, or have them practice make ten addition or subtraction on the back of their papers. Collect and correct any Practice Sets completed within the allotted time.

Standards Check: Time (5 minutes)

Materials: (T/S) Personal white board, time recording sheet (Fluency Template)

Note: This review fluency provides an opportunity to monitor which students can tell and write time in hours and half hours. When students draw hands for times to the half hour, make sure the hour hand is approximately halfway between the numbers.

- T: Draw hands on the template's analog clock to show times to the hour and half hour.
- S: (Write the time on the digital clock, and fill in the appropriate sentence frame.)
- T: Write times to the hour and half hour on the digital clock.
- S: (Draw the hands on the analog clock, and fill in the appropriate sentence frame.)

Lesson 26:   Solve compare with bigger or smaller unknown problem types.   315

Lesson 26

Fluency Favorite or Standards Review (5 minutes)

Note: If needed, repeat one of the Standards Check fluency activities. If not, select a class favorite fluency activity, or begin the Concept Development.

Concept Development (35 minutes)

Materials: (S) Problem Set

Note: By working with double tape diagrams as related to the varying comparison problem types, students have a way to approach any comparison problem.

- How do we set up our story as a tape diagram?
- Read carefully and determine who has more.
- Is every part of the story represented in your tape diagram?

Suggested Delivery of Instruction for Solving Word Problems

1. Model the problem, calculate, and write a statement.

Choose two pairs of students to work on chart paper while the others work independently or in pairs at their seats. Review the following questions before beginning the first problem:

- How do we set up our story into a double tape diagram?
- Read carefully. Who has more?
- Is every part of the story represented in your tape diagram?

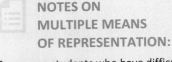

NOTES ON MULTIPLE MEANS OF REPRESENTATION:

Encourage students who have difficulty moving to the tape diagram representation as the position of the unknown changes to draw a number bond as part of their work. Some students more easily relate to the tape diagram through its similarities with number bonds.

As students work, circulate and support. Some students may feel stuck and struggle with choosing the appropriate method to use. Encourage and support them in learning to persevere and make sense of the problems.

After three minutes, have the two pairs of students share their labeled diagrams. Allow students to briefly question their peers until it is agreed the diagrams or drawings represent the story correctly. Then, give everyone two to three minutes to finish work on that question. All students should write their equations and statements of the answer.

2. Assess the solution for reasonableness.

Give all students one to two minutes to assess and explain the reasonableness of their solution to a partner. For about one minute, have the demonstrating students receive and respond to feedback and questions from their peers.

Lesson 26: Solve compare with bigger or smaller unknown problem types.

3. As a class, notice the ways the drawing depicts the story and the solution.

Ask questions to help students recognize how each part of their diagram matches the story and solution. This helps students begin to see how the same process can help them solve varying word problems. Keep at least one chart-paper sample of each solution for reference later in the lesson.

Problem 1 (*Compare with difference unknown.*)

Tony is reading a book with 16 pages. Maria is reading a book that has 10 pages. How much longer is Tony's book than Maria's book?

Note: After students have explained their tape diagram and solution accurately, point to sections of the tape diagram, and ask the class questions such as, "What does this part represent? How do you know?"

For the next five problems, have only the students at the board share their work so that students have time to work through and discuss all six problems. Choose one or two probing questions that support student development, as needed.

Problem 2 (*Compare with bigger unknown.*)

Shanika built a block tower using 14 blocks. Tamra built a tower by using 5 more blocks than Shanika. How many blocks did Tamra use to build her tower?

Note: For many children, Problem 2 is more challenging to solve than Problem 1 because one of the sets being compared (Tamra's) has a missing part. Some students may quickly find an accurate solution from adding the two numbers (14 + 5) but may not demonstrate understanding in their drawing.

> **NOTES ON MULTIPLE MEANS OF ACTION AND EXPRESSION:**
>
> If students do not have experience with a context such as the one used in Problem 2, act out the problem with a few student volunteers before having the class begin to draw and solve the problem.

Problem 3 (*Compare with difference unknown.*)

Darnel walked for 10 minutes to get to Kiana's house. The next day, Kiana took a shortcut and walked to Darnel's house in 8 minutes. How much shorter in time was Kiana's walk?

Note: Problem 3 brings students back to a *compare with difference unknown* problem type, which they should be gaining confidence in solving. Celebrate the strategies students use to achieve such successes as a motivator to continue persevering at problems they initially find challenging.

Problem 4 (*Compare with smaller unknown.*)

Lee read 16 pages in a book. Kim read 4 fewer pages in her book. How many pages did Kim read?

Note: Students sometimes struggle with the term *fewer*, making Problem 4 more challenging. Using relatively small differences (such as 4) can support students in visualizing the problem and learning the vocabulary.

Problem 5 (*Compare with bigger unknown. More or fewer suggesting the incorrect operation.*)

Nikil's soccer team has 13 players. Nikil has 4 fewer players on his team than Rose's team. How many players are on Rose's team?

A STORY OF UNITS

Lesson 26 1•6

Note: Problem 5 is challenging because *fewer than* suggests the incorrect operation. Similar to Problem 4, the small difference between the two team sizes (4 players) is intentionally selected to support students in working with this challenging problem type.

Problem 6 (*Compare with smaller unknown. More or fewer suggesting the incorrect operation.*)

After dinner, Darnel washed 15 spoons. He washed 9 more spoons than forks. How many forks did Darnel wash?

Note: Problem 6 uses *more than*, but students must subtract to find the number of forks that were washed. As a final problem, notice that the difference between the two sets being compared is 9, a much larger difference than used in the previous two problems.

Problem Set (10 minutes)

Students should do their personal best to complete the Problem Set within the allotted 10 minutes. For some classes, it may be appropriate to modify the assignment by specifying which problems they work on first. Some problems do not specify a method for solving. Students should solve these problems using the RDW approach used for Application Problems.

Student Debrief (10 minutes)

Lesson Objective: Solve *compare with bigger or smaller unknown* problem types.

The Student Debrief is intended to invite reflection and active processing of the total lesson experience.

Invite students to review their solutions for the Problem Set. They should check work by comparing answers with a partner before going over answers as a class. Look for misconceptions or misunderstandings that can be addressed in the Debrief. Guide students in a conversation to debrief the Problem Set and process the lesson.

Any combination of the questions below may be used to lead the discussion.

- Look at Problem 1. What did you draw? How did your drawing help you solve the problem?

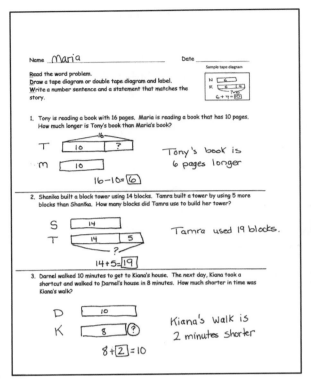

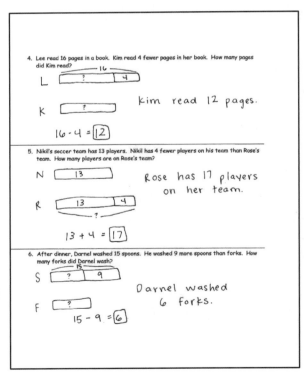

318 | Lesson 26: Solve compare with bigger or smaller unknown problem types.

- Look at Problem 3. How is your drawing similar or different from the drawing you made for Problem 1?
- Look at Problem 4. How was setting up your drawing similar to Problem 5? Explain your thinking.
- Why is it important to read the stories carefully? When you see the words *more than,* does it always mean you have to add to find your solution? Use examples from your Problem Set to support your thinking.

Exit Ticket (3 minutes)

After the Student Debrief, instruct students to complete the Exit Ticket. A review of their work will help with assessing students' understanding of the concepts that were presented in today's lesson and planning more effectively for future lessons. The questions may be read aloud to the students.

Lesson 26: Solve compare with bigger or smaller unknown problem types.

Name _____ Date _____

Read the word problem.
Draw a tape diagram or double tape diagram and label.
Write a number sentence and a statement that matches the story.

Sample Tape Diagram

N | 6
R | 6 | 4
  ? = 10
6 + 4 = 10

1. Tony is reading a book with 16 pages. Maria is reading a book that has 10 pages. How much longer is Tony's book than Maria's book?

2. Shanika built a block tower using 14 blocks. Tamra built a tower by using 5 more blocks than Shanika. How many blocks did Tamra use to build her tower?

3. Darnel walked 10 minutes to get to Kiana's house. The next day, Kiana took a shortcut and walked to Darnel's house in 8 minutes. How much shorter in time was Kiana's walk?

Lesson 26: Solve compare with bigger or smaller unknown problem types.

4. Lee read 16 pages in a book. Kim read 4 fewer pages in her book. How many pages did Kim read?

5. Nikil's soccer team has 13 players. Nikil has 4 fewer players on his team than Rose's team. How many players are on Rose's team?

6. After dinner, Darnel washed 15 spoons. He washed 9 more spoons than forks. How many forks did Darnel wash?

Lesson 26: Solve compare with bigger or smaller unknown problem types.

Name _____   Date _____

Read the word problem.
Draw a tape diagram or double tape diagram and label.
Write a number sentence and a statement that matches the story.

Sample Tape Diagram

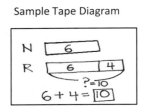

Maria jumped off the diving board into the pool 3 fewer times than Emi. Maria jumped off the diving board 14 times. How many times did Emi jump off the diving board?

A STORY OF UNITS **Lesson 26 Homework 1•6**

Name _____ Date _____

Read the word problem.
Draw a tape diagram or double tape diagram and label.
Write a number sentence and a statement that matches the story.

Sample tape diagram

1. Fatima walks 15 blocks home from school. Ben walks 8 blocks. How much longer is Fatima's walk home from school than Ben's?

2. Maria bought a basket with 13 strawberries in it. Darnel bought a basket with 4 more strawberries than Maria. How many strawberries did Darnel's basket have in it?

3. Tamra has 5 books checked out from the library. Kim has 11 books checked out from the library. How many fewer books does Tamra have checked out than Kim?

Lesson 26: Solve compare with bigger or smaller unknown problem types.

A STORY OF UNITS   Lesson 26 Homework 1•6

4. Kiana picked 12 apples from the tree. She picked 6 fewer apples than Willie. How many apples did Willie pick from the tree?

5. During recess, Emi found 16 rocks. She found 5 more rocks than Peter. How many rocks did Peter find?

6. The first grade football team has 12 players. The first grade team has 6 fewer players than the second grade team. How many players are on the second grade team?

A STORY OF UNITS

Lesson 26 Fluency Template   1•6

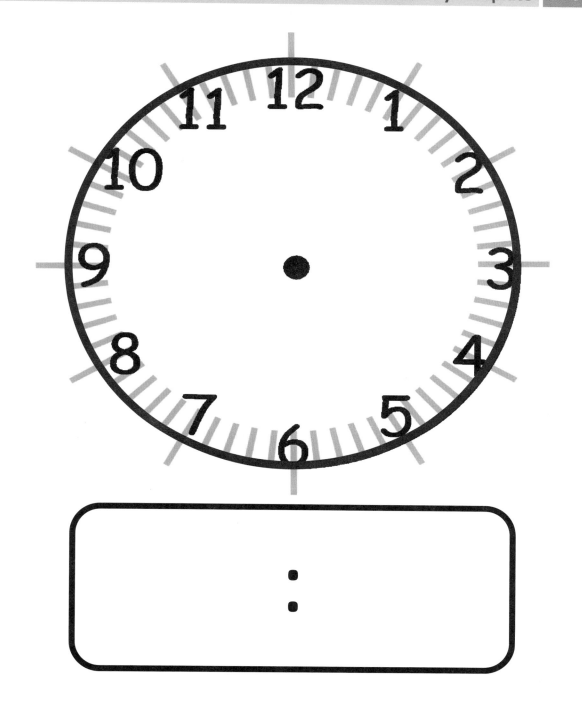

It is ___ o'clock.   It is half past ___.

_____

time recording sheet

Lesson 26: Solve compare with bigger or smaller unknown problem types.

| A STORY OF UNITS | Lesson 27 1•6 |

Lesson 27

Objective: Share and critique peer strategies for solving problems of varied types.

Suggested Lesson Structure

- ■ Fluency Practice   (13 minutes)
- ■ Concept Development (37 minutes)
- ■ Student Debrief   (10 minutes)
- **Total Time**   **(60 minutes)**

Fluency Practice (13 minutes)

- Core Fluency Differentiated Practice Sets **1.OA.6**   (5 minutes)
- Standards Check: Shapes **1.G.1, 1.G.2**   (8 minutes)

Core Fluency Differentiated Practice Sets (5 minutes)

Materials: (S) Core Fluency Practice Sets (Lesson 1 Core Fluency Practice Sets)

Note: Give the appropriate Practice Set to each student. Help students become aware of their improvement. After students do today's Practice Sets, ask them to stand if they tried a new level today or improved their score from the previous day. Consider having students clap once for each person standing to celebrate improvement.

Students complete as many problems as they can in 90 seconds. Assign a counting pattern and start number for early finishers, or have them practice make ten addition or subtraction on the back of their papers. Collect and correct any Practice Sets completed within the allotted time.

Standards Check: Shapes (8 minutes)

Materials: (T) Two-dimensional shape flashcards (Fluency Template 1), three-dimensional objects used in Module 5 Lesson 3 (S) Personal white board, shapes recording sheet (Fluency Template 2)

Note: This activity reviews the attributes and names of two-dimensional and three-dimensional shapes. Remember that a square is also a rectangle and a rhombus, and a cube is also a rectangular prism.

1. Invite students to look at their templates and to read the names of the two-dimensional shapes and attributes with the teacher. Show a shape card or object. Students circle the name(s) of the shape and complete the attributes section. Repeat for all two-dimensional shapes.

326   Lesson 27:   Share and critique peer strategies for solving problems of varied types.

A STORY OF UNITS
Lesson 27 1•6

2. Invite students to look at their templates and to read the names of the three-dimensional shapes and attributes with the teacher. Show a three-dimensional object. Students circle the name(s) of the shape and complete the attributes section. Repeat for all three-dimensional shapes.

3. Show two- or three-dimensional shapes. Ask students to circle the other shapes that could be used, if any, to create them.

Concept Development (37 minutes)

Materials: (T) Chart paper (S) Problem Set

Students sit at the tables next to their partner with their Problem Sets.

Note: In today's lesson, students work on their Problem Set and solve the varied problem types they encountered throughout the year. Selected pairs of students then discuss their methods for solving the problems and explain their work. After they share, the whole class participates in a discussion as students make comments and suggestions and ask each other questions.

- How does your work or tape diagram help you solve the problem?
- A compliment I could give you is...?
- A question I have for you is...?
- One way you might improve your work would be...?
- Let's look for similarities and differences in our drawings and strategies.

Suggested Delivery of Instruction for Sharing and Critiquing Peer Strategies

1. Solve varied problem types using the RDW process.

For each story problem, invite two pairs of students to model their work on chart paper while the others work independently or in pairs. Choose new pairs for each problem, and consider selecting students who use varied strategies for solving.

As students work, circulate and provide support. Some students may feel stuck and struggle with picking the appropriate method or choosing between a single or a double tape diagram to use. Encourage and support them in learning to persevere to make sense of the problems.

NOTES ON MULTIPLE MEANS OF ENGAGEMENT:

Observe levels of student understanding, and select the most appropriate problem type to focus on during today's Concept Development.

2. In partnerships, share and critique peer strategies.

Give students one to two minutes to explain their methods of solving and how they found their solution with their partners or with another pair of students.

Lesson 27: Share and critique peer strategies for solving problems of varied types.

A STORY OF UNITS   **Lesson 27   1•6**

3. **As a class and with partners, share and critique peer strategies.**

For Problems 1 and 2, share and critique peer strategies as a class. For about one minute, have the demonstrating students share their methods and explain their work. The rest of the class may raise questions, and the presenters respond to feedback and questions from their peers. For the remaining problems, have students share and critique with their partners using the chart with question frames. Finally, all students return to their work and make improvements.

Problem 1 (*Add to with change unknown.*)

Nine letters came in the mail on Monday. Some more letters were delivered on Tuesday. Then, there were 13 letters. How many letters were delivered on Tuesday?

Note: Students have worked with this problem type throughout the year. Some students may use addition to solve, while others use subtraction. It is important to see that different operations can be used as long as the story problem has been analyzed accurately.

NOTES ON MULTIPLE MEANS OF ACTION AND EXPRESSION:

If students struggle with computation, use smaller numbers or numbers that are close together so they can focus on how to interpret and solve different problem types.

Problem 2 (*Take apart with addend unknown.*)

Ben and Tamra found a total of 18 seeds in their watermelon slices. Ben found 7 seeds in his slice. How many seeds did Tamra find?

Note: Like Problem 1, students may solve using addition or subtraction. Larger numbers are used within the problem, which may also promote conversation about place value as students discuss their solution strategies.

Problem 3 (*Add to with start unknown.*)

Some children were playing on the playground. Eight children came to join, and now there are 14 children. How many children were on the playground in the beginning?

Note: Problem 3 is challenging because it begins with an unknown. If both members of a partnership are struggling, remind them to read the story one sentence at a time and check that their drawing represents each sentence. Students might use concrete manipulatives and then draw after they understand the relationships within the problem.

Problem 4 (*Compare with difference unknown.*)

Willie walked for 7 minutes. Peter walked for 14 minutes. How much shorter in time was Willie's walk?

Note: This problem challenges students to notice that they are working with a comparison problem type.

Problem 5 (*Compare with bigger unknown.*)

Emi saw 12 ants walking in a row. Fran saw 6 more ants than Emi. How many ants did Fran see?

Note: Students must recognize that the second sentence in this story problem only gives part of the necessary information to determine how many ants Fran saw. Support students with questions such as, "Who are the characters? Who saw more ants? What can you draw?"

A STORY OF UNITS — Lesson 27 1•6

Problem 6 (*Compare with smaller unknown.*)

Shanika has 13 cents in her front pocket. She has 8 fewer cents in her back pocket. How many cents does Shanika have in her back pocket?

Note: Problem 6 presents some of the same challenges as Problem 5, this time using the term *fewer*. Support students with questions such as, "Are you comparing, or are you putting together? What are you comparing? What can you draw?"

Problem Set (10 minutes)

Students should do their personal best to complete the Problem Set within the allotted 10 minutes. For some classes, it may be appropriate to modify the assignment by specifying which problems they work on first. Some problems do not specify a method for solving. Students should solve these problems using the RDW approach used for Application Problems.

Student Debrief (10 minutes)

Lesson Objective: Share and critique peer strategies for solving problems of varied types.

The Student Debrief is intended to invite reflection and active processing of the total lesson experience.

Invite students to review their solutions for the Problem Set. They should check work by comparing answers with a partner before going over answers as a class. Look for misconceptions or misunderstandings that can be addressed in the Debrief. Guide students in a conversation to debrief the Problem Set and process the lesson.

Any combination of the questions below may be used to lead the discussion.

- Which problems did you and your partner find challenging today? How did your discussion help you to solve the problem or to improve your strategies for solving the problem?

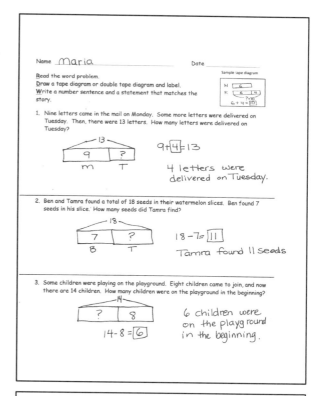

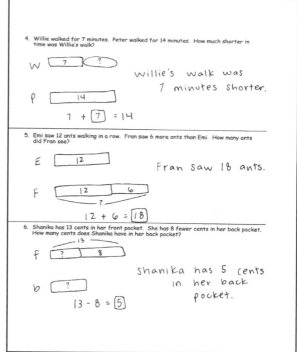

Lesson 27: Share and critique peer strategies for solving problems of varied types.

- What were some of the similarities in the way you and your partner drew and solved the problems? What were some of the differences?
- How did seeing your partner's work help improve your own work? Show your improvement to the class.
- What compliments did you give your partner about her work? Show the class an example of your partner's work.

Exit Ticket (3 minutes)

After the Student Debrief, instruct students to complete the Exit Ticket. A review of their work will help with assessing students' understanding of the concepts that were presented in today's lesson and planning more effectively for future lessons. The questions may be read aloud to the students.

A STORY OF UNITS

Lesson 27 Problem Set 1•6

Name _____ Date _____

Read the word problem.
Draw a tape diagram or double tape diagram and label.
Write a number sentence and a statement that matches the story.

Sample Tape Diagram

N	6	
R	6	4

?=10
6 + 4 = 10

1. Nine letters came in the mail on Monday. Some more letters were delivered on Tuesday. Then, there were 13 letters. How many letters were delivered on Tuesday?

2. Ben and Tamra found a total of 18 seeds in their watermelon slices. Ben found 7 seeds in his slice. How many seeds did Tamra find?

3. Some children were playing on the playground. Eight children came to join, and now there are 14 children. How many children were on the playground in the beginning?

Lesson 27: Share and critique peer strategies for solving problems of varied types.

4. Willie walked for 7 minutes. Peter walked for 14 minutes. How much shorter in time was Willie's walk?

5. Emi saw 12 ants walking in a row. Fran saw 6 more ants than Emi. How many ants did Fran see?

6. Shanika has 13 cents in her front pocket. She has 8 fewer cents in her back pocket. How many cents does Shanika have in her back pocket?

Lesson 27 Exit Ticket 1•6

Name _____ Date _____

Read the word problem.
Draw a tape diagram or double tape diagram and label.
Write a number sentence and a statement that matches the story.

Sample Tape Diagram

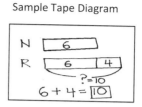

Emi tried on 8 fewer costumes than Nikil. Emi tried on 4 costumes. How many costumes did Nikil try on?

A STORY OF UNITS **Lesson 27 Homework 1•6**

Name _____ Date _____

Read the word problem.
Draw a tape diagram or double tape diagram and label.
Write a number sentence and a statement that matches the story.

Sample Tape Diagram

1. Eight students lined up to go to art. Some more lined up to go to music. Then, there were 12 students in line. How many students lined up to go to music?

2. Peter rode his bike 5 blocks. Rose rode her bike 13 blocks. How much shorter was Peter's ride?

3. Lee and Anton collected 16 leaves on their walk. Nine of the leaves were Lee's. How many leaves were Anton's?

4. The team counted 11 soccer balls inside the net. They counted 5 fewer soccer balls outside of the net. How many soccer balls were outside of the net?

5. Julio saw 14 cars drive by his house. Julio saw 6 more cars than Shanika. How many cars did Shanika see?

6. Some students were eating lunch. Four students joined them. Now, there are 17 students eating lunch. How many students were eating lunch in the beginning?

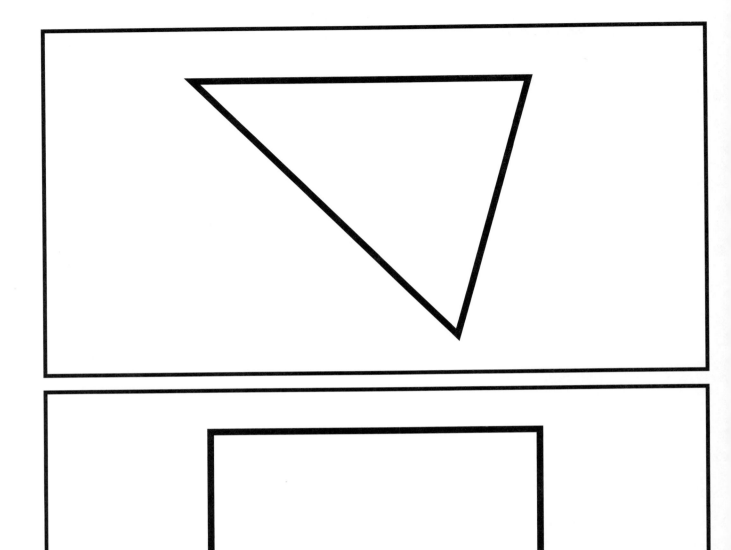

two-dimensional shape flashcards

Lesson 27 Fluency Template 1

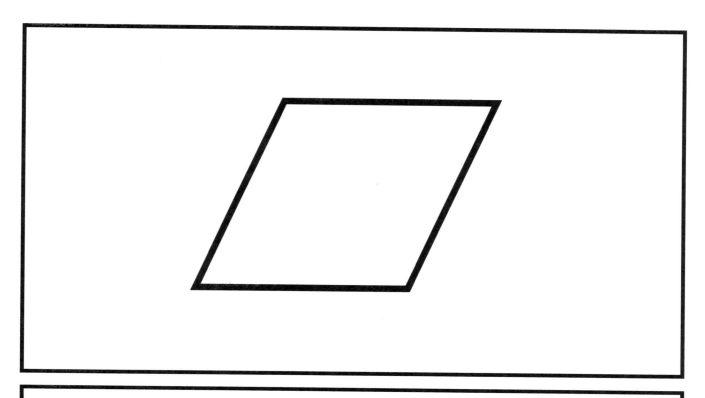

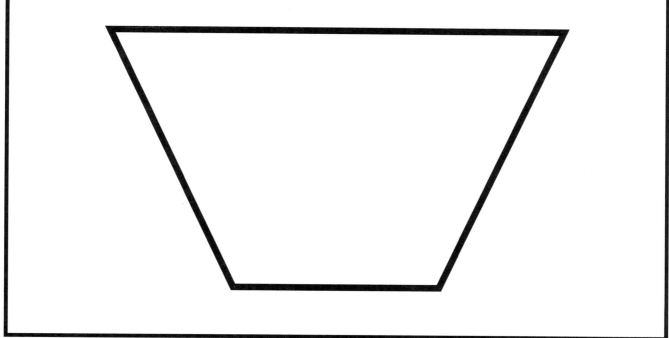

two-dimensional shape flashcards

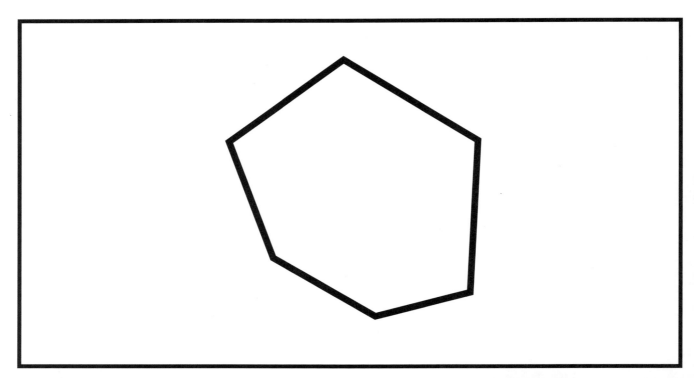
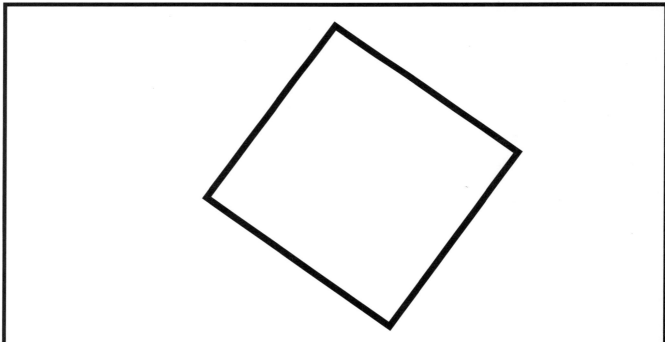

two-dimensional shape flashcards

Lesson 27 Fluency Template 1

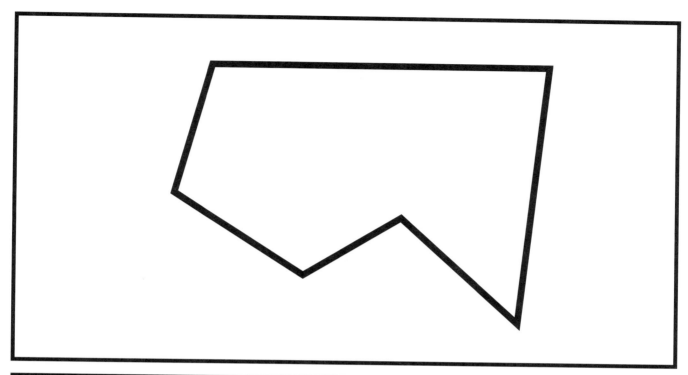

two-dimensional shape flashcards

Lesson 27: Share and critique peer strategies for solving problems of varied types.

2-D SHAPES	3-D SHAPES
circle	sphere
triangle	cone
rectangle	cylinder
rhombus	rectangular prism
square	cube
trapezoid	
hexagon	

____ corners   ____ corners

____ square corners   ____ faces

____ sides   ____ straight edges

Are all sides the same length?   Are all faces the same shape?

yes   no   yes   no

shapes recording sheet

A STORY OF UNITS

End-of-Module Assessment Task 1•6

Name _____ Date _____

1. Use the RDW process to solve the following problems. Write the statement on the line.

   a. Tamra has 12 coins. Willie has 8 coins. How many more coins does Tamra have than Willie?

   _____.

   b. 16 coins are on the table. 11 of them are pennies, and the rest are dimes. How many dimes are there?

   _____.

   c. Peter has 6 fewer coins than Nikil. Nikil has 9 coins. How many coins does Peter have?

   _____.

Module 6: Place Value, Comparison, Addition and Subtraction to 100

341

A STORY OF UNITS

End-of-Module Assessment Task 1•6

2. Fill in the missing numbers in each sequence:

   a. 115, 116, _____, _____, _____, 120

   b. _____, 101, _____, 99, _____

3. Use the word bank to write the number and value of each coin.

Coin Names		Coin Values	
nickel	dime	1 cent	5 cents
quarter	penny	10 cents	25 cents

    _____   _____

   _____   _____

   _____   _____

   _____   _____

342   Module 6:   Place Value, Comparison, Addition and Subtraction to 100

4. Mark says that 87 is the same as 7 tens 17 ones. Suki says that 87 is the same as 8 tens 7 ones. Are they correct? Explain your thinking.

5. Use <, =, or > to compare the pairs of numbers.

   a. 6 tens ◯ 42 ones

   b. 69 ◯ 75

   c. 75 ◯ 6 tens 15 ones

   d. 8 tens 14 ones ◯ 7 tens 4 ones

6. Find the mystery numbers. Explain how you know the answers.

   a. 10 more than 89 is _____.

   b. 10 less than 89 is _____.

   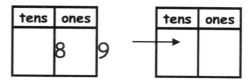

   c. 1 more than 89 is _____.

   d. 1 less than 89 is _____.

   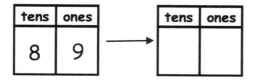

7. Solve for each unknown number. Use the space provided to draw quick tens, a number bond, or the arrow way to show your work. You may use your kit of ten-sticks if needed.

a. 90 + 3 = _____	b. 50 + 40 = _____	c. 80 − 30 = _____
d. 100 − _____ = 40	e. 78 + 6 = _____	f. 47 + 40 = _____
g. 65 + 34 = _____	h. 75 + 25 = _____	i. 47 + 36 = _____

End-of-Module Assessment Task
Standards Addressed — Topics A–F

Represent and solve problems involving addition and subtraction.

1.OA.1 Use addition and subtraction within 20 to solve word problems involving situations of adding to, taking from, putting together, taking apart, and comparing, with unknowns in all positions, e.g., by using objects, drawings, and equations with a symbol for the unknown number to represent the problem. (See CCSS-M Glossary, Table 1.)

Extend the counting sequence.

1.NBT.1 Count to 120, starting at any number less than 120. In this range, read and write numerals and represent a number of objects with a written numeral.

Understand place value.

1.NBT.2 Understand that the two digits of a two-digit number represent amounts of tens and ones. Understand the following as special cases:

   a. 10 can be thought of as a bundle of ten ones—called a "ten."

   c. The numbers 10, 20, 30, 40, 50, 60, 70, 80, 90 refer to one, two, three, four, five, six, seven, eight, or nine tens (and 0 ones).

1.NBT.3 Compare two two-digit numbers based on meanings of the tens and ones digits, recording the results of comparisons with the symbols >, =, and <.

Use place value understanding and properties of operations to add and subtract.

1.NBT.4 Add within 100, including adding a two-digit number and a one-digit number, and adding a two-digit number and a multiple of 10, using concrete models or drawings and strategies based on place value, properties of operations, and/or the relationship between addition and subtraction; relate the strategy to a written method and explain the reasoning used. Understand that in adding two-digit numbers, one adds tens and tens, ones and ones; and sometimes it is necessary to compose a ten.

1.NBT.5 Given a two-digit number, mentally find 10 more or 10 less than the number, without having to count; explain the reasoning used.

1.NBT.6 Subtract multiples of 10 in the range 10–90 from multiples of 10 in the range 10–90 (positive or zero differences), using concrete modules or drawings and strategies based on place value, properties of operations, and/or the relationship between addition and subtraction; relate the strategy to a written method and explain the reasoning used.

Tell and write time and money.[1]

1.MD.3 Tell and write time in hours and half-hours using analog and digital clocks. Recognize and identify coins, their names, and their value.

---

[1] Focus on money.

A STORY OF UNITS

End-of-Module Assessment Task 1•6

Evaluating Student Learning Outcomes

A Progression Toward Mastery is provided to describe steps that illuminate the gradually increasing understandings that students develop *on their way to proficiency.* In this chart, this progress is presented from left (Step 1) to right (Step 4). The learning goal for students is to achieve Step 4 mastery. These steps are meant to help teachers and students identify and celebrate what the students CAN do now and what they need to work on next.

A Progression Toward Mastery				
Assessment Task Item	STEP 1 Little evidence of reasoning without a correct answer. (1 Point)	STEP 2 Evidence of some reasoning without a correct answer. (2 Points)	STEP 3 Evidence of some reasoning with a correct answer or evidence of solid reasoning with an incorrect answer. (3 Points)	STEP 4 Evidence of solid reasoning with a correct answer. (4 Points)

1
 	
1.OA.1
	Student's answers are incorrect, and there is no evidence of reasoning.	Student's answers are incorrect, but there is evidence of reasoning. For example, the student is able to write a number sentence.	Student's answers are correct, but the responses are incomplete (e.g., may be missing labels for the drawing, an addition sentence, or an explanation). The student's work is essentially strong.	Student correctly:
				• Solves each word problem.
				  a. Tamra has 4 more coins than Willie.
				  b. There are 5 dimes on the table.
				  c. Peter has 3 coins.
				• Demonstrates understanding of the problem situation through drawing/modeling.

2 1.NBT.1	Student is unable to complete any sequence of numbers.	Student completes at least part of one sequence.	Student completes at least one sequence as well as at least one number in the additional sequence.	Student identifies all numbers in the sequences: • 115, 116, **117**, **118**, **119**, 120 • **102**, 101, **100**, 99, 98

346 Module 6: Place Value, Comparison, Addition and Subtraction to 100

End-of-Module Assessment Task

A Progression Toward Mastery

3 **1.MD.3**	Student is unable to match more than two coins with either the proper name or the proper value.	Student accurately matches at least three elements within the set but mixes the value or the names for more than one pair of coins.	Student accurately matches one set of coin information with another set but mixes either the value or the name of one pair of coins.	Student correctly matches the image, name, and value of each coin: • Dime, 10 cents • Penny, 1 cent • Nickel, 5 cents • Quarter, 25 cents
4 **1.NBT.2**	Student demonstrates little to no understanding of comparing numbers based on tens and ones, answering incorrectly. There is no evidence of reasoning.	Student uses drawings or words to accurately depict at least one of the two numbers, demonstrating limited understanding of the use of place value to compare numbers.	Student demonstrates some understanding of using place value to compare numbers but does not fully explain reasoning using place value. OR Student answers incorrectly because of an error such as transcription but demonstrates strong understanding of place value through drawing or words.	Student correctly uses drawings or words that depict place value to accurately explain that 87 is the same as both 7 tens 17 ones and 8 tens 7 ones.
5 **1.NBT.2** **1.NBT.3**	Student is unable to use symbols to compare numbers and is unable to correctly answer any of the four comparisons.	Student has limited ability to use symbols to compare numbers, correctly answering one of the four comparisons.	Student has some ability to use symbols to compare numbers, correctly answering two or three of the four comparisons.	Student answers: a. > b. < c. = d. >

Module 6: Place Value, Comparison, Addition and Subtraction to 100

A Story of Units — End-of-Module Assessment Task 1•6

A Progression Toward Mastery

6 **1.NBT.5**	Student demonstrates little or no understanding of mentally adding or subtracting 10. Answers are incorrect, and there is no evidence of reasoning.	Student demonstrates limited understanding of mentally adding or subtracting 10, identifying at least two correct mystery numbers but does not complete any charts accurately.	Student demonstrates the ability to mentally add or subtract 10, correctly identifying four mystery numbers but reasoning is unclear because no charts have been completed accurately. OR Student accurately completes the charts but makes an error in mental calculation on one or two of (a), (b), (c), or (d.)	Student identifies 99, 79, 90, and 88, and accurately completes the charts to depict the arrow way.

7
 	
1.NBT.4
1.NBT.6
	Student answers two or fewer questions correctly.	Student answers at least three of nine correctly and demonstrates misunderstandings of place value.	Student answers at least six of nine correctly or uses a sound process throughout with calculation errors.	Student correctly:
				- Solves
				  a. 93
				  b. 90
				  c. 50
				  d. 60
				  e. 84
				  f. 87
				  g. 99
				  h. 100
				  i. 83
				- Represents the process to accurately solve through drawings, number bonds, or the arrow way. The notation demonstrates the use of a sound strategy for adding or subtracting.

Module 6: Place Value, Comparison, Addition and Subtraction to 100

Name Maria   Date

1. Use the RDW process to solve the following problems. Write the statement on the line.

   a. Tamra has 12 coins. Willie has 8 coins. How many more coins does Tamra have than Willie?

   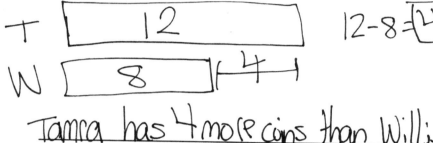

   Tamra has 4 more coins than Willie.

   b. 16 coins are on the table. 11 of them are pennies and the rest are dimes. How many dimes are there?

   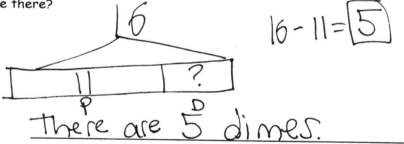

   There are 5 dimes.

   c. Peter has 6 fewer coins than Nikil. Nikil has 9 coins. How many coins does Peter have?

   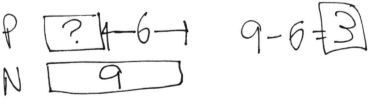

   Peter has 3 coins.

End-of-Module Assessment Task 1•6

2. Fill in the missing numbers in each sequence:

   a. 115, 116, __117__, __118__, __119__, 120

   b. __102__, 101, __100__, 99, __98__

3. Use the word bank to write the number and value of each coin.

Coin Names				Coin Values	
nickel	dime	quarter	penny	1 cent	5 cents
				10 cents	25 cents

   dime   10 cents

   penny   1 cent

   nickel   5 cents

   quarter   25 cents

4. Mark says that 87 is the same as 7 tens 17 ones. Suki says that 87 is the same as 8 tens 7 ones. Are they correct? Explain your thinking.

Mark and Suki are both right.

8 tens 7 ones is 80 + 7 = 87

7 tens 17 ones is 70 + 17 = 87

87 = 87

5. Use <, =, or > to compare the pairs of numbers.

   a. 6 tens ⟩ 42 ones

   b. 69 ⟨ 75

   c. 75 = 6 tens 15 ones

   d. 8 tens 14 ones ⟩ 7 tens 4 ones

6. Find the mystery numbers. Explain how you know the answers.

a. 10 more than 89 is __99__

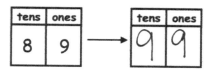

b. 10 less than 89 is __79__

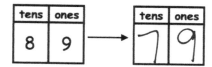

c. 1 more than 89 is __90__

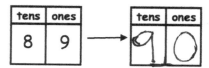

d. 1 less than 89 is __88__

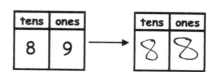

7. Solve for each unknown number. Use the space provided to draw quick tens, a number bond, or the arrow way to show your work. You may use your kit of ten-sticks if needed.

a. 90 + 3 = __93__ 90 →(+3) 93	b. 50 + 40 = __90__ ||||| ||||	c. 80 − 30 = __50__ ||||| (⊠)

d. 100 − __60__ = 40 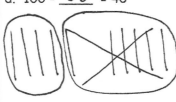	e. 78 + 6 = __84__
	   ∧
	  2 4
	78 + 2 = 80
	80 + 4 = 84
		f. 47 + 40 = __87__ 

g. 65 + 34 = __99__
   ∧
  30 4
65 + 30 = 95
95 + 4 = 99
	h. 75 + 25 = __100__
	   ∧
	  20 5
	75 + 5 = 80
	80 + 20 = 100
		i. 47 + 36 = __83__
		   ∧
		  3 33
		47 + 3 = 50
		50 + 33 = 83

A STORY OF UNITS

Mathematics Curriculum

GRADE 1 • MODULE 6

Topic G
Culminating Experiences

Focus Standard:	Topic G is a celebration of students' learning over the course of the year. Focus Standards are not applicable.		
Instructional Days:	3		
Coherence	-Links from:	G1–M4	Place Value, Comparison, Addition and Subtraction to 40
	-Links to:	G2–M3	Place Value, Counting, and Comparison of Numbers to 1,000

Topic G culminates not only Module 6, but also a full year of learning for Grade 1 students. It is a joyous celebration of the great progress of all students. During each lesson, students recognize how much they know now in comparison with the start of the year. They celebrate this learning by using their acquired skills and knowledge to enjoy entertaining games and activities with their peers.

During Lessons 28 and 29, students play games with cards and dice that celebrate their progress in fluently adding and subtracting within 10 and 20. All of the games are played with materials that students can find at home or bring home from school to encourage engaging summer practice.

To culminate the year, students create folder covers that can be used to bring home the math work from the year. The covers are designed to illustrate students' learning across the course of the year and to celebrate their individual accomplishments.

A Teaching Sequence Toward Mastery of Culminating Experiences
Objective 1: Celebrate progress in fluency with adding and subtracting within 10 (and 20). Organize engaging summer practice. (Lessons 28–29)
Objective 2: Create folder covers for work to be taken home illustrating the year's learning. (Lesson 30)

Topic G: Culminating Experiences

A STORY OF UNITS

Lesson 28 1•6

Lesson 28

Objective: Celebrate progress in fluency with adding and subtracting within 10 (and 20). Organize engaging summer practice.

Suggested Lesson Structure

- Fluency Practice   (10 minutes)
- Application Problem   (5 minutes)
- Culminating Activity   (35 minutes)
- Student Debrief   (10 minutes)
- **Total Time**   **(60 minutes)**

Fluency Practice (10 minutes)

- Sprint: Count Dots **K.CC.5**   (10 minutes)

Sprint: Count Dots (10 minutes)

Materials: (S) Count Dots Sprint

Note: This Sprint is the one students completed on the first day of school. Repeating it in the final days of school will likely bring students joy as they recognize the ease with which they are able to do it after a year of mathematical growth. Be sure to assign a counting sequence for early finishers!

Application Problem (5 minutes)

Darnel answered 30 problems on Side B of his Count Dots Sprint today. He was proud because he answered 20 more problems today than he did on the first day of school. How many problems did he answer on the first day of school?

Note: This *compare with smaller unknown* problem challenges students by suggesting the wrong operation.

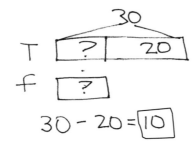

Darnel answered 10 problems on the first day.

354   Lesson 28:   Celebrate progress in fluency with adding and subtracting within 10 (and 20). Organize engaging summer practice.

A STORY OF UNITS   Lesson 28 1•6

Culminating Activity (35 minutes)

Materials: (T) Organizational chart for center assignments (example to the right)
(S) Numeral cards (Template 1), Target Practice (Template 2), Race to the Top (Template 3), personal white board, die

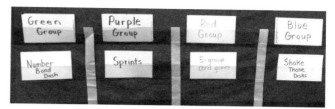

Note: In the next two lessons, students revisit some of their favorite fluency activities from the year to celebrate and reflect on their progress.

Take the steps listed below to prepare for the culminating activity:

- Choose from the suggested activities, or select other fluency favorites based on the needs and interests of the class.
- Prepare materials and stations.
- On the Problem Set, before making copies for today's lesson, write the names of the activities selected. (See the picture to the right.) This is an opportunity for students to reflect on their progress.

Note: Students work with these centers again tomorrow as a host to guests who might be parents, support teachers, or kindergarten buddies.

T: Today, we are going to celebrate our fluency progress. Think about the fluency activities we did this year. Which were your favorites?

T: How did they help you improve your counting, adding, and subtracting skills? Share your ideas with your partner.

S: Happy Counting helped me count forward and backward. → Sprints helped me with addition and subtraction facts. → Coin drops helped with counting on.

T: Great! Today, I have some of those activities set up at centers. You will start at one center and rotate at my signal to the other centers. Review instructions for each center and assign partners. Students spend about five minutes at each center.

NOTES ON MULTIPLE MEANS OF ENGAGEMENT:

It is important to provide students with the math tools they need to play these games successfully. Support students with the use of manipulatives and possibly their personal white boards.

Lesson 28: Celebrate progress in fluency with adding and subtracting within 10 (and 20). Organize engaging summer practice.

355

A STORY OF UNITS Lesson 28 1•6

Choose from the fluency celebration centers suggested below. Set up the number of centers that works best for the class.

Missing Part: Make Ten

Materials: (S) Numeral cards (Template 1)

Each partner holds a card up to his or her forehead. The partner tells how many more are needed to make ten. Students must guess the cards on their foreheads. Partners can play simultaneously, each putting a card to his or her forehead.

Target Practice

Materials: (S) Personal white boards with Target Practice (Template 2), die per pair

Follow the directions on the game board.

Race to the Top

Materials: (S) Personal white boards with Race to the Top (Template 3), 2 dice per pair

Partners take turns rolling the dice, saying an addition sentence and recording the sums on the graph. The game ends when time runs out or one of the columns reaches the top of the graph.

Subtraction with Cards

Materials: (S) Numeral cards (Template 1)

Partners combine their numeral cards and place them face down between them.

- Each partner flips over two cards and subtracts the smaller number from the larger one.
- The partner with the smallest difference keeps the cards played by both players in that round.
- If the differences are equal, the cards are set aside, and the winner of the next round keeps the cards from both rounds.
- When all of the cards have been used, the player with the most cards wins.

Number Bond Addition and Subtraction

Materials: (S) Personal white boards, die per pair

Allow partners to choose a number less than 20 for their whole and roll the die to determine one of the parts.

- Both students write two addition and two subtraction sentences with a square for the unknown number in each equation and solve for the missing number.
- They then exchange boards and check each other's work.

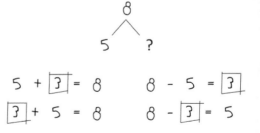

Lesson 28: Celebrate progress in fluency with adding and subtracting within 10 (and 20). Organize engaging summer practice.

Make Ten Addition and Take from Ten Subtraction with Partners

Materials: (S) Personal white boards

Partners alternate practicing the make ten and take from ten strategies.

Make Ten Addition:

- Partners choose an addend for each other from 1 to 10.
- On their personal boards, students add their number to 9, 8, and 7. Remind students to write the two addition sentences they learned in Module 2.
- Partners then exchange boards and check each other's work.

```
9 + 5 = 14      8 + 5 = 13      7 + 5 = 12
   ∧               ∧               ∧
  1 4             2 3             3 2

9 + 1 = 10      8 + 2 = 10      7 + 3 = 10
10 + 4 = 14     10 + 3 = 13     10 + 2 = 12
```

Take from Ten Subtraction:

- Partners choose a minuend for each other between 10 and 20.
- On their personal white boards, students subtract 9, 8, and 7 from their number. Remind students to write the two number sentences (e.g., to solve 13 – 8, they write 10 – 8 = 2, 2 + 3 = 5).
- Partners then exchange boards and check each other's work.

```
13 - 9 = 4      13 - 8 = 5      13 - 7 = 6
   ∧               ∧               ∧
 10  3           10  3           10  3

10 - 9 = 1      10 - 8 = 2      10 - 7 = 3
 1 + 3 = 4       2 + 3 = 5       3 + 3 = 6
```

Analogous Addition Sentences

Materials: (S) Personal white board, dice

- Step 1: Each partner rolls a die and writes the number rolled. They then make a list, adding 1 ten to their number on each new line up to 3 tens. (See the diagram to the right.)
- Step 2: Students write equations, adding the number on their partners' die to each line.
- Partners exchange boards and check each other's work.

Note: This game can be modified by using dice that have more than 6 sides. Students should be ready to add numbers to 20 and add multiples of 10 to these numbers.

STEP 1	
Partner A	Partner B
4	3
14	13
24	23
34	33

STEP 2	
Partner A	Partner B
4 + 3 = 7	3 + 4 = 7
14 + 3 = 17	13 + 4 = 17
24 + 3 = 27	23 + 4 = 27
34 + 3 = 37	33 + 4 = 37

Lesson 28: Celebrate progress in fluency with adding and subtracting within 10 (and 20). Organize engaging summer practice.

Student Debrief (10 minutes)

Lesson Objective: Celebrate progress in fluency with adding and subtracting within 10 (and 20). Organize engaging summer practice.

The Student Debrief is intended to invite reflection and active processing of the total lesson experience.

Invite students to review their center work today. They should reflect on their work with a partner before sharing as a class. Guide students in a conversation to debrief the centers and reflect on their learning.

Any combination of the questions below may be used to lead the discussion.

- What is something you did today that you could not do before you came to first grade?
- Which of today's centers seemed easy? How does your experience today compare with the first time you did them?
- Are there any activities that were still a little challenging? What might you do to get better?
- Which of these games might be fun to play over the summer so you can keep your math skills sharp?

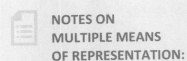

NOTES ON MULTIPLE MEANS OF REPRESENTATION:

Teachers should feel a sense of pride as they see their students demonstrate strategies to make math easy. It is also exciting when students are able to explain how they are thinking and relate concepts to one another.

Exit Ticket

Note: There is no Exit Ticket for this lesson.

A

Name _____ Date _____

Number Correct: _____

*Write the number of dots. Try to find ways to group the dots to make counting easier!

1. ●●		16. ●●●●● ●●●●	
2. ●●●		17. ●●●●● ●●●	
3. ●●●●		18. ●●●●● ●●●●●	
4. ●●●		19. ●●●●● ●●	
5. ●		20. ●●●●● ●	
6. ●●●●		21. ●●●●● ●●●	
7. ●●●●●		22. ●●●●● ●●●●●	
8. ●●●●		23. ●●●● ●●●●●	
9. ●●●●● ●		24. ●●●●● ●●●	
10. ●●●●● ●●		25. ●●● ●● ●●●●●	
11. ●●●●●		26. ●●●●● ●●	
12. ●●●●		27. ●●● ●● ●● ●●●	
13. ●●●●● ●		28. ●● ●● ●● ●●	
14. ●●●●● ●●●		29. ●● ●●● ● ●●	
15. ●●●●● ●●		30. ●● ●● ●● ●●	

Lesson 28: Celebrate progress in fluency with adding and subtracting within 10 (and 20). Organize engaging summer practice.

359

Number Correct:

B

Name ______________________ Date ____________

*Write the number of dots. Try to find ways to group the dots to make counting easier!

1.	●		16.	●●●●● ●●●	
2.	●●		17.	●●●●● ●●●●	
3.	●		18.	●●●●● ●●	
4.	●●●●		19.	●●●●● ●●●	
5.	●●●		20.	●●●●● ●●●●●	
6.	●●●●●		21.	●●●●● ●●●●	
7.	●●●●		22.	●●●●● ●●●●●	
8.	●●●●●		23.	● ●●●● ●●●●●	
9.	●●●●● ●●		24.	●●●●● ●●●●●	
10.	●●●●● ●		25.	●● ●●●●●	
11.	●●●●● ●●●		26.	●●● ● ●● ●●	
12.	●●●●● ●		27.	●● ●●● ●●● ●●	
13.	●●●●●		28.	●● ● ● ● ●● ●●	
14.	●●●●● ●●		29.	●● ●● ● ●● ●●	
15.	●●●●● ●		30.	●● ● ●●●● ●●	

EUREKA MATH™

A STORY OF UNITS

Lesson 28 Problem Set 1•6

Name _____ Date _____

1. Circle the smiley face that shows your level of fluency for each activity.

Activity	I still need some practice.	I can complete, but I still have some questions.	I am fluent.
a.			
b.			
c.			
d.			
e.			
f.			

2. Which activity helped you the most in becoming fluent with your facts to 10?

Lesson 28: Celebrate progress in fluency with adding and subtracting within 10 (and 20). Organize engaging summer practice.

A STORY OF UNITS **Lesson 28 Homework 1•6**

Name _____   Date _____

1. Teach a family member some of our counting activities. Check all the activities you do together.

   ☐ Happy Count by ones.
   ☐ Happy Count by tens.
   ☐ Count by ones the Say Ten Way.
   ☐ Count by tens the Say Ten Way. First, start at 0; then, start at 7.
   ☐ Movement counting—count while doing squats, arm rolls, jumping jacks, etc.

2. Write the numbers from 91 to 120:

91		93							
				105					
								119	

3. Count backward by tens from 97 to 7.

   97, _____, 77, _____, _____, _____, _____, _____, _____, _____.

4. On the back of your paper, write as many sums and differences within 20 as you can. Circle the ones that were hard for you at the beginning of the year!

0	1	2	3
4	5	<u>6</u>	7
8	<u>9</u>	10	10
	10	5	5

numeral cards

Target Practice

Target Number:

Choose a *target number* between 6 and 10, and write it in the middle of the circle on the top of the page. Roll a die. Write the number rolled in the circle at the end of one of the arrows. Then, make a bull's-eye by writing the number needed to make your target in the other circle.

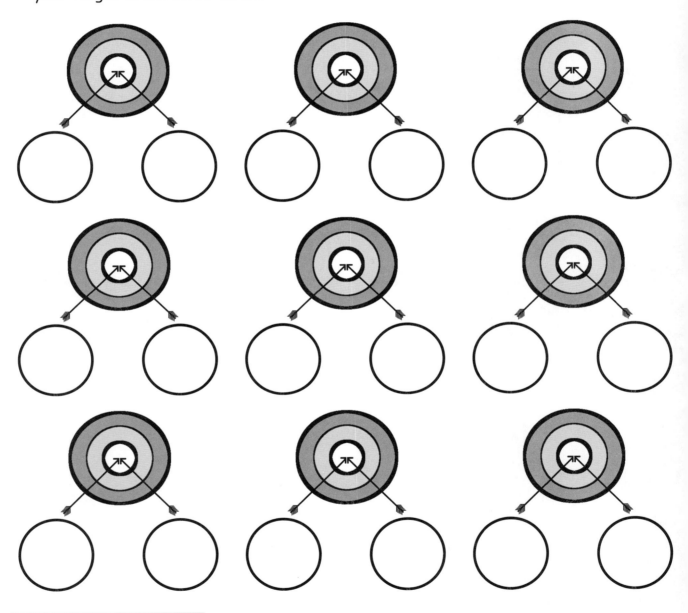

target practice

Name _____ Date _____

 Race to the Top!

2	3	4	5	6	7	8	9	10	11	12

race to the top

Lesson 29

Objective: Celebrate progress in fluency with adding and subtracting within 10 (and 20). Organize engaging summer practice.

Suggested Lesson Structure

- Fluency Practice (3 minutes)
- Application Problem (5 minutes)
- Culminating Activity (42 minutes)
- Student Debrief (10 minutes)
- **Total Time** **(60 minutes)**

Fluency Practice (3 minutes)

- Number Bond Dash: 10 **1.OA.6**   (3 minutes)

Number Bond Dash: 10 (3 minutes)

Materials: (S) Number Bond Dash: 10 (Pattern Sheet)

Note: In Module 1, students used the Number Bond Dash to build fluency with decompositions to 10. Doing it today will likely bring students joy as they realize the ease with which they complete an activity that was once a challenge.

Application Problem (5 minutes)

In October, Tamra's best score on the Number Bond Dash was 15 problems. Today, she correctly answered 10 more problems. What was Tamra's score today?

Note: This *add to with result unknown* problem ties into today's Fluency Celebration. Students likely relate to Tamra because they have just recognized their own improvement on the Number Bond Dash.

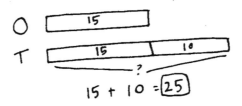

A STORY OF UNITS

Lesson 29 1•6

Culminating Activity (42 minutes)

Materials: (S) Various fluency activities for center work

Note: Choose one of the following two options.

1. Invite parents, buddies from a kindergarten class, support staff, or another audience to the Fluency Celebration. Set up the same fluency centers selected yesterday. Empower students to teach activities to their guests. Students can either host one station or travel with a group as a guide.

2. Replace some of yesterday's centers with different suggested Fluency Celebration centers or other fluency favorites based on the needs and interests of the class.

   T: Welcome to our Fluency Celebration. Today the class will show you some of the fluency activities we have worked on this year.

Circulate as students teach the fluency games to their invited guests.

> **NOTES ON MULTIPLE MEANS OF REPRESENTATION:**
>
> During the Fluency Celebration, be sure to assign partners suitably matched for the games played. Some partners are better when matched by ability, and others may work better with one stronger student.

Student Debrief (10 minutes)

Lesson Objective: Celebrate progress in fluency with adding and subtracting within 10 (and 20). Organize engaging summer practice.

The Student Debrief is intended to invite reflection and active processing of the total lesson experience.

Invite students to review their center work today. They should reflect on their work with a partner before sharing as a class. Guide students in a conversation to debrief the centers and reflect on their learning.

Any combination of the questions below may be used to lead the discussion.

> **NOTES ON MULTIPLE MEANS OF ACTION AND EXPRESSION:**
>
> Giving students an opportunity to share and teach their favorite games empowers them at the end of their year. Celebrate English language learners as they use the language they have been learning in class all year to explain their thinking.

- What is something you did today that you could not do before you came to first grade?
- What did you do to teach your guests the activities? Which ones were more difficult for them?
- Why do you think everyone says that when you teach something to someone else, you remember it much better?
- Are there any activities that were still a little challenging? What might you do to get better?
- Which of these games might be fun to play over the summer so you can keep your math skills sharp?

Lesson 29: Celebrate progress in fluency with adding and subtracting within 10 (and 20). Organize engaging summer practice.

A STORY OF UNITS   Lesson 29 Pattern Sheet   1•6

Name _____   Date _____

Number Bond Dash!

Directions: Do as many as you can in 90 seconds.
Write the amount you finished here:

1. 10 / 10, ☐
2. 10 / 9, ☐
3. 10 / 8, ☐
4. 10 / 9, ☐
5. 10 / 10, ☐
6. 10 / ☐, 9
7. 10 / ☐, 8
8. 10 / ☐, 7
9. 10 / ☐, 8
10. 10 / ☐, 7
11. 10 / 6, ☐
12. 10 / 7, ☐
13. 10 / 6, ☐
14. 10 / 5, ☐
15. 10 / 4, ☐
16. 10 / ☐, 6
17. 10 / ☐, 4
18. 10 / ☐, 3
19. 10 / ☐, 4
20. 10 / ☐, 3
21. 10 / 0, ☐
22. 10 / 1, ☐
23. 10 / 2, ☐
24. 10 / 4, ☐
25. 10 / 2, ☐

Lesson 29: Celebrate progress in fluency with adding and subtracting within 10 (and 20). Organize engaging summer practice.

A STORY OF UNITS   Lesson 30 1•6

Lesson 30

Objective: Create folder covers for work to be taken home illustrating the year's learning.

Suggested Lesson Structure

■ Culminating Activity   (50 minutes)
■ Student Debrief   (10 minutes)
    Total Time   **(60 minutes)**

Note: Today is intended to be an opportunity for closure and reflection. There is no Fluency Practice or Application Problem today.

Culminating Activity (50 minutes)

Materials: (T) White pocket folder (S) White pocket folders, crayons, colored pencils or markers, 2 envelopes

Note: If white pocket folders are not available, manila file folders or 18" × 24" white paper can be used to make folders.

- T: What are some of the math concepts we learned this year?
- S: Addition and subtraction!
- T: What can we draw on our folders to remember these concepts?
- S: Number sentences! → 5-group drawings! → Number bonds!
- T: (Model a drawing that represents addition and subtraction. Circulate as students represent addition and subtraction on their folders.)
- T: What are some other concepts we have learned?
- S: Tens and ones!
- T: What can we draw on our folders to represent tens and ones?
- S: Dimes and pennies! → Place value charts! → Quick tens and ones! → Tens-sticks and cubes! → Adding where we lined up tens with tens and ones with ones!

Continue this sequence of questions and prompts as students review the important concepts they have learned throughout Grade 1. These should include measurement, data, three-dimensional shapes, two-dimensional shapes, and word problems. When the folders are complete, they can be used to send home summer packet directions, supplies, and completed work from Grade 1 that represents student learning.

Lesson 30:   Create folder covers for work to be taken home illustrating the year's learning.

Summer Packet Should Include:

- Lesson 30 Summer Packet. (Summer Packet is found at the end of this lesson.)
- Single-sided numeral or 5-group cards. (Consider sending home the set used by the student during the school year or a template to cut new cards from Lesson 28.)
- 5 Core Fluency Sprints. (Other Grade 1 Sprints may also be selected based on the needs of students.)
- Core Fluency Differentiated Practice Sets.

Student Debrief (10 minutes)

Lesson Objective: Create folder covers for work to be taken home illustrating the year's learning.

Invite students to review their work today. They should reflect on their learning throughout the year by sharing their illustration with a partner before sharing as a class. Guide students in a conversation to debrief their reflections.

Any combination of the questions below may be used to lead the discussion.

- What drawing did you make to represent addition and subtraction? Why did you choose this drawing?
- How did you show that addition and subtraction are related?
- What shapes did you draw?
- As I circulated, I saw lots of drawings: number bonds, place value charts, tape diagrams, 5-groups, and quick tens. How do you think some of these pictures have helped you to understand math this year?
- How did you show your learning about word problems?
- What do you think you are going to learn next year in second grade?

Lesson 30 Summer Packet

Name _____   Date _____

Complete a math activity each day. Color the box for each day you do the suggested activity.

Summer Math Review: Weeks 1-5

	Monday	Tuesday	Wednesday	Thursday	Friday
Week 1	Count from 87 to 120 and back.	Play Addition with Cards.	Use your tangram pieces to make a Fourth of July picture.	Use quick tens and ones to draw 76.	Complete a Sprint.
Week 2	Do counting squats. Count from 45 to 60 and back the Say Ten Way.	Play Subtraction with Cards.	Make a graph of the types of fruits in your kitchen. What did you find out from your graph?	Solve 36 + 57. Draw a picture to show your thinking.	Complete a Sprint.
Week 3	Write numbers from 37 to as high as you can in one minute, while whisper-counting the Say Ten Way.	Play Target Practice or Shake Those Disks for 9 and 10.	Measure a table with spoons and then with forks. Which did you need more of? Why?	Use real coins or draw coins to show as many ways to make 25 cents as you can.	Complete a Sprint.
Week 4	Do jumping jacks as you count up by tens to 120 and back down to 0.	Play Race and Roll Addition or Addition with Cards.	Go on a shape scavenger hunt. Find as many rectangles or rectangular prisms as you can.	Use quick tens and ones to draw 45 and 54. Circle the greater number.	Complete a Sprint.
Week 5	Write the numbers from 75 to 120.	Play Race and Roll Subtraction or Subtraction with Cards.	Measure the route from your bathroom to your bedroom. Walk heel to toe, and count your steps.	Add 5 tens to 23. Add 2. What number did you find?	Complete a Sprint.

Lesson 30: Create folder covers for work to be taken home illustrating the year's learning.

A STORY OF UNITS

Lesson 30 Summer Packet 1•6

Name _____ Date _____

Complete a math activity each day. Color the box for each day you do the suggested activity.

Summer Math Review: Weeks 6-10

	Monday	Tuesday	Wednesday	Thursday	Friday
Week 6	Count by ones from 112 to 82. Then, count from 82 to 112.	Play Missing Part for 7.	Write a story problem for 9 + 4.	Solve 64 + 38. Draw a picture to show your thinking.	Complete a Core Fluency Practice Set.
Week 7	Do counting squats. Count down from 99 to 75 and back up the Say Ten Way.	Play Race and Roll Addition or Addition with Cards.	Graph the colors of all your pants. What did you find out from your graph?	Draw 14 cents with dimes and pennies. Draw 10 more cents. What coins did you use?	Complete a Core Fluency Practice Set.
Week 8	Write the numbers from 116 to as low as you can in one minute.	Play Missing Part for 8.	Write a story problem for 7 + ___ = 12.	Use quick tens and ones to draw 76. Draw dimes and pennies to show 59 cents.	Complete a Core Fluency Practice Set.
Week 9	Do jumping jacks as you count up by tens from 9 to 119 and back down to 9.	Play Race and Roll Subtraction or Subtraction with Cards.	Go on a shape scavenger hunt. Find as many circles or spheres as you can.	Use quick tens and ones to draw 89 and 84. Circle the number that is less.	Complete a Core Fluency Practice Set.
Week 10	Write numbers from 82 to as high as you can in one minute, while whisper counting the Say Ten Way.	Play Target Practice or Shake Those Disks for 6 and 7.	Measure the steps from your bedroom to the kitchen, walking heel to toe, and then have a family member do the same thing. Compare.	Solve 47 + 24. Draw a picture to show your thinking.	Complete a Core Fluency Practice Set.

Lesson 30: Create folder covers for work to be taken home illustrating the year's learning.

Addition (or Subtraction) with Cards

Materials: 2 sets of numeral cards 0–10

- Shuffle the cards, and place them face down between the two players.
- Each partner flips over two cards and adds them together or subtracts the smaller number from the larger one.
- The partner with the largest sum or smallest difference keeps the cards played by both players in that round.
- If the sums or differences are equal, the cards are set aside, and the winner of the next round keeps the cards from both rounds.
- When all the cards have been used, the player with the most cards wins.

Sprint

Materials: Sprint (Sides A and B)

- Do as many problems on Side A as you can in one minute. Then, try to see if you can improve your score by answering even more of the problems on Side B in a minute.

Target Practice

Materials: 1 die

- Choose a target number to practice (e.g., 10).
- Roll the die, and say the other number needed to hit the target. For example, if you roll 6, say 4, because 6 and 4 make ten.

Shake Those Disks

Materials: Pennies

The amount of pennies needed depends on the number being practiced. For example, if students are practicing sums for 10, they need 10 pennies.

- Shake your pennies, and drop them on the table.
- Say two addition sentences that add together the heads and tails. (For example, if they see 7 heads and 3 tails, they would say 7 + 3 = 10 and 3 + 7 = 10.)
- Challenge: Say four addition sentences instead of two. (For example, 10 = 7 + 3, 10 = 3 + 7, 7 + 3 = 10, and 3 + 7 = 10.)

Race and Roll Addition (or Subtraction)

Materials: 1 die

Addition

- Both players start at 0.
- They each roll a die and then say a number sentence adding the number rolled to their total. (For example, if a player's first roll is 5, the player says 0 + 5 = 5.)
- They continue rapidly rolling and saying number sentences until someone gets to 20 without going over. (For example, if a player is at 18 and rolls 5, the player would continue rolling until she gets a 2.)
- The first player to 20 wins.

Subtraction

- Both players start at 20.
- They each roll a die and then say a number sentence subtracting the number rolled from their total. (For example, if a player's first roll is 5, the player says 20 − 5 = 15.)
- They continue rapidly rolling and saying number sentences until someone gets to 0 without going over. (For example, if a player is at 5 and rolls 6, the player would continue rolling until she gets a 5.)
- The first player to 0 wins.

Answer Key

Eureka Math
Grade 1
Module 6

Special thanks go to the Gordon A. Cain Center and to the Department of Mathematics at Louisiana State University for their support in the development of *Eureka Math*.

Published by the non-profit Great Minds

Copyright © 2015 Great Minds. No part of this work may be reproduced, sold, or commercialized, in whole or in part, without written permission from Great Minds. Non-commercial use is licensed pursuant to a Creative Commons Attribution-NonCommercial-ShareAlike 4.0 license; for more information, go to http://greatminds.net/maps/math/copyright. "Great Minds" and "Eureka Math" are registered trademarks of Great Minds.

Printed in the U.S.A.
This book may be purchased from the publisher at eureka-math.org
10 9 8 7 6 5 4 3 2

A STORY OF UNITS

GRADE 1 — Mathematics Curriculum

GRADE 1 • MODULE 6

Answer Key
GRADE 1 • MODULE 6
Place Value, Comparison, Addition and Subtraction to 100

Module 6: Place Value, Comparison, Addition and Subtraction to 100

Lesson 1

Core Fluency Practice

Set A

1. 6
2. 6
3. 6
4. 6
5. 7
6. 7
7. 8
8. 7
9. 7
10. 6
11. 8
12. 8
13. 6
14. 7
15. 8
16. 9
17. 10
18. 9
19. 9
20. 10
21. 8
22. 9
23. 10
24. 10
25. 8
26. 7
27. 10
28. 9
29. 10
30. 9

Set B

1. 0
2. 6
3. 1
4. 2
5. 7
6. 1
7. 6
8. 1
9. 7
10. 2
11. 3
12. 4
13. 5
14. 4
15. 2
16. 3
17. 4
18. 2
19. 3
20. 4
21. 3
22. 4
23. 5
24. 6
25. 7
26. 8
27. 7
28. 6
29. 5
30. 6

A STORY OF UNITS

Lesson 1 Answer Key 1•6

Set C

1. 1
2. 5
3. 5
4. 1
5. 9
6. 1
7. 5
8. 5
9. 2
10. 2
11. 3
12. 3
13. 2
14. 2
15. 3
16. 3
17. 2
18. 2
19. 3
20. 3
21. 4
22. 4
23. 3
24. 3
25. 4
26. 4
27. 3
28. 3
29. 3
30. 3

Set D

1. 6
2. 5
3. 6
4. 7
5. 4
6. 5
7. 7
8. 0
9. 1
10. 3
11. 3
12. 4
13. 6
14. 2
15. 4
16. 6
17. 5
18. 1
19. 2
20. 2
21. 4
22. 5
23. 3
24. 4
25. 5
26. 4
27. 3
28. 2
29. 3
30. 2

Lesson 1 Answer Key 1•6

Set E

1. 6
2. 4
3. 3
4. 7
5. 2
6. 7
7. 3
8. 4
9. 9
10. 5
11. 4
12. 4
13. 2
14. 2
15. 2
16. 3
17. 4
18. 2
19. 6
20. 2
21. 3
22. 5
23. 3
24. 7
25. 2
26. 7
27. 2
28. 3
29. 5
30. 3

Problem Set

1. 6
2. 6
3. 5
4. 6

Exit Ticket

5

Homework

1. 3
2. 10
3. 6
4. 4

Lesson 2

Problem Set

1. 8
2. 9
3. 4
4. 6
5. 20
6. 6

Exit Ticket

8

Homework

1. 5
2. 9
3. 11
4. 15
5. 10
6. 17

Lesson 3

Problem Set

1. 4, 3; 4, 3
2. 8, 6; 86, 8, 6
3. 7, 8; 78
4. 8, 7; 87
5. 9, 6; 96
6. 10, 0; 100
7. 7, 3; 73
8. 5, 4; 54
9.
   a. 4, 0
   b. 4, 6
   c. 59
   d. 95
   e. 7, 5
   f. 7, 0
   g. 6, 0
   h. 80
   i. 55
   j. 100

Exit Ticket

1. 8, 9; 89
2.
   a. 9, 0
   b. 87

Homework

1. 5, 2; 5, 2
2. 9, 8; 98, 9, 8
3. 9, 7; 97
4. 5, 9; 59
5. 10, 0; 100
6. 8, 6; 86
7. 6, 7; 67
8. 7, 5; 75
9.
   a. 7, 0
   b. 7, 6
   c. 49
   d. 94
   e. 6, 5
   f. 6, 0
   g. 9, 0
   h. 100
   i. 83
   j. 80

Lesson 4

Problem Set

1. 40, 3, 43; 43; 43
2. 40, 6, 46; 46; 46
3. 50, 7, 57; 50; 7; 57
4. 70, 5, 75; 70, 5; 75
5. 60, 8, 68; 60, 8, 68; 6, 8, 68
6. 90, 2, 92; 90, 2, 92; 9, 2, 92
7. 7, 4; 74, 70, 4; 4, 7, 74
8. 8, 6; 86, 80, 6; 6, 8, 86
9. 7, 4; 70, 4, 74; 7, 4, 74
10. 10; 100, 0, 100; 10, 0, 100
11. a. 56
    b. 80
    c. 6
    d. 89

Exit Ticket

1. 8, 9; 80, 9, 89; 8, 9, 89
2. a. 92
   b. 9

Homework

1. 70, 6, 76; 76; 76
2. 40, 5, 45; 45; 45
3. 60, 9, 69; 60, 9; 69
4. 90, 7, 97; 90, 7; 97
5. 80, 4, 84; 80, 4, 84; 8, 4, 84
6. 50, 8, 58; 50, 8, 58; 5, 8, 58
7. 5, 6; 56, 50, 6; 6, 5, 56
8. 6, 8; 68, 60, 8; 8, 6, 68
9. 7, 5; 70, 5, 75; 7, 5, 75
10. 9; 90, 0, 90; 9, 0, 90
11. a. 86
    b. 50
    c. 5
    d. 84

Lesson 5

Problem Set

1. a. 69; 1 cube drawn
    b. 78; 1 ten-stick drawn
    c. 61; 1 ten-stick crossed off
    d. 69; 1 cube crossed off

2. a. 69; 6, 9
    b. 58; 5, 9, $\xrightarrow{-1}$, 5, 8
    c. 60; 5, 9, $\xrightarrow{+1}$, 6, 0
    d. 49; 5, 9, $\xrightarrow{-10}$, 4, 9

3. a. 11
    b. 71
    c. 77
    d. 80
    e. 100

4. a. 20
    b. 70
    c. 71
    d. 88
    e. 100

5. a. 11
    b. 51
    c. 50
    d. 79
    e. 99

6. a. 10
    b. 50
    c. 64
    d. 71
    e. 90

7. a. 43
    b. 86
    c. 70
    d. 64
    e. 70
    f. 50
    g. 75
    h. 79
    i. 100
    j. 87; 67

Exit Ticket

1. a. 68; 6, 9, $\xrightarrow{-1}$, 6, 8
    b. 79; 6, 9, $\xrightarrow{+10}$, 7, 9

2. a. 41
    b. 87
    c. 90

3. a. 60
    b. 72
    c. 100

4. a. 74
    b. 69
    c. 99

5. a. 70
    b. 89
    c. 90

Homework

1. a. 89; 1 ten-stick drawn
    b. 71; 1 ten-stick crossed off
    c. 80; 1 cube drawn
    d. 79; 1 cube crossed off
2. a. 85; 8, 5
    b. 76; 7, 5, $\xrightarrow{+1}$, 7, 6
    c. 78; 8, 8, $\xrightarrow{-10}$, 7, 8
    d. 87; 8, 8, $\xrightarrow{-1}$, 8, 7
3. a. 41
    b. 51
    c. 66
    d. 70
    e. 100
4. a. 70
    b. 80
    c. 87
    d. 99
    e. 100
5. a. 52
    b. 72
    c. 70
    d. 79
    e. 99
6. a. 40
    b. 50
    c. 74
    d. 81
    e. 90
7. a. 53
    b. 76
    c. 60
    d. 84
    e. 90
    f. 70
    g. 77
    h. 69
    i. 100
    j. 94; 74

Lesson 6

Problem Set

1. a. <
   b. <
   c. >
   d. >
   e. <
   f. >
   g. =
   h. =

2. a. Is equal to; 29 = 29
   b. Is less than; 79 < 80
   c. Is greater than; 100 > 10
   d. Is less than; 61 < 66

3. a. <
   b. >
   c. =
   d. =
   e. =
   f. =
   g. =
   h. <

Exit Ticket

a. Is less than; 36 < 63
b. Is greater than; 90 > 89
c. Is equal to; 52 = 52
d. Is less than; 42 < 44

Homework

1. a. >
   b. <
   c. >
   d. =
   e. <
   f. <
   g. =
   h. <

2. a. Is greater than; 42 > 12
   b. Is equal to; 67 = 67
   c. Is less than; 37 < 73
   d. Is greater than; 34 > 24
   e. Is less than; 59 < 95

Lesson 7

Problem Set

1. a. 72; 74; 75; 78
    b. 85; 88; 90
    c. 92; 95; 98
    d. 101; 103; 105; 107; 108
    e. 112; 115; 118; 120
2. 98, 99, 100, 101, 102, 103, 104, 105, 106, 107, 108, 109, 110, 111, 112, 113, 114, 115, 116, 117, 118, 119, 120
3. Sequence (a) circled; 107, 108, 109, 110, 111
4. a. 117, 118, 119
    b. 116, 117, 119
    c. 102, 103
    d. 99, 100, 101, 102

Exit Ticket

1. a. 89
    b. 98; 100
    c. 109; 110
    d. 118; 120
2. a. 118; 120
    b. 110, 111, 112

Homework

1. a. 72; 73; 75; 76; 77; 78
    b. 81; 83; 84; 86; 88; 89
    c. 92; 94; 95; 97; 98; 100
    d. 101; 103; 104; 106; 107; 109
    e. 112; 113; 115; 117; 118; 120
2. 100, 102, 103, 104, 105, 106, 107, 108, 109, 110, 111, 112, 113, 114, 115, 116, 117, 118, 119, 120
3. Sequence (b) circled; 96, 97, 98, 99, 100, 101
4. a. 115, 116, 117
    b. 117, 118, 119
    c. 103, 104, 105
    d. 90, 91, 92, 93

Lesson 8

Problem Set

1. a. 7, 4
    b. 7, 8
    c. 91
    d. 109
    e. 11, 6
    f. 10, 3
    g. 112
    h. 120
    i. 105
    j. 10, 2

2. a. 9 tens 7 ones
    b. 10 tens 7 ones
    c. 110
    d. 10 tens 5 ones
    e. 101
    f. 12 tens 0 ones
    g. 11 tens 8 ones

Exit Ticket

1. a. 8, 3
    b. 94
    c. 115
    d. 10, 6

2. a. 102
    b. 114

Homework

1. a. 8, 1
    b. 9, 8
    c. 117
    d. 108
    e. 10, 4
    f. 11, 1

2. a. 92
    b. 84
    c. 113
    d. 109
    e. 101
    f. 116

3. a. 102
    b. 9 tens 5 ones
    c. 11 tens 4 ones
    d. 11 tens 0 ones
    e. 108
    f. 10 tens 0 ones
    g. 11 tens 8 ones

Lesson 9

Sprint

Side A

1.	6	11.	28	21.	39
2.	16	12.	18	22.	29
3.	26	13.	8	23.	19
4.	15	14.	19	24.	1
5.	25	15.	29	25.	10
6.	35	16.	39	26.	10
7.	7	17.	10	27.	1
8.	17	18.	20	28.	24
9.	27	19.	30	29.	24
10.	37	20.	40	30.	34

Side B

1.	5	11.	27	21.	39
2.	15	12.	17	22.	29
3.	25	13.	7	23.	19
4.	16	14.	18	24.	1
5.	26	15.	28	25.	10
6.	36	16.	38	26.	10
7.	6	17.	10	27.	1
8.	16	18.	20	28.	29
9.	26	19.	30	29.	29
10.	36	20.	40	30.	39

A STORY OF UNITS — Lesson 9 Answer Key 1•6

Problem Set

1. 9, 8; 98
2. 10, 8; 108
3. 11, 8; 118
4. 10, 5; 105
5. 11, 6; 116
6. 11, 9; 119
7. 12, 0; 120
8. 109; 10 quick tens and 9 ones drawn
9. 120; 12 quick tens drawn

Exit Ticket

1. 11, 8; 118
2. a. 110; 11 quick tens drawn
   b. 101; 10 quick tens and 1 one drawn

Homework

1. 9, 6; 96
2. 10, 6; 106
3. 11, 6; 116
4. 10, 9; 109
5. 12, 0; 120
6. 10, 7; 107
7. 11, 0; 110
8. 110; 11 quick tens drawn
9. 105; 10 quick tens and 5 ones drawn

Lesson 10

Recording Sheet

Answers will vary.

Problem Set

1. 2, 5; 50
2. 70, 30, 40; 3, 4, 7; 30 + 40 = 70
3. 80, 50, 30; 8, 3, 5; 80 − 30 = 50
4. 90, 60, 30; 6, 3, 9; 60 + 30 = 90
5. 90, 70, 20; 9, 2, 7; 90 − 20 = 70
6. 60
7. 60 − 20 = 40
8. 70 + 30 = 100
9. 100 − 30 = 70
10. 90 − 40 = 50
11. a. 80
    b. 20
    c. 60
    d. 60
    e. 80
    f. 20
    g. 90
    h. 70
    i. 30

Exit Ticket

1. a. 90
   b. 20
   c. 40
2. 80 − 30 = 50

Homework

1. a. Line drawn to number bond 60, 40, __; 20
   b. Line drawn to number bond 90, 30, __; 60
   c. Line drawn to number sentence 80 − __ = 60; 20
   d. Line drawn to number sentence __ − 40 = 60; 100
2. a. 60
   b. 70 − 20 = 50
   c. 70 + 30 = 100
   d. 60 − 40 = 20
3. a. 20
   b. 50
   c. 80
   d. 90
   e. 50
   f. 40
   g. 80
   h. 70
   i. 50

Lesson 11

Problem Set

1. Number bond drawn; 54 + 20 = 74
2. Number bond drawn; 54 + 40 = 94
3. Number bond drawn; 38 + 50 = 88
4. Number bond drawn; 19 + 80 = 99
5.
   a. 87
   b. 87
   c. 65
   d. 85
   e. 93
   f. 79

6.
   a. 52
   b. 98
   c. 50
   d. 26

Exit Ticket

a. 92; quick ten drawing or number bond drawn
b. 87; quick ten drawing or number bond drawn

Homework

1.
   a. 52 + 10 = 62
   b. 50 + 34 = 84
   c. 26 + 30 = 56
   d. 30 + 48 = 78
2.
   a. 78; number bond drawn
   b. 84; number bond drawn
   c. 86; number bond drawn
   d. 87; number bond drawn
   e. 88; number bond drawn
   f. 95; number bond drawn

3.
   a. 92
   b. 98
   c. 50
   d. 47

Lesson 12

Problem Set

1. a. 96
    b. 97
    c. 79
    d. 100
    e. 100
    f. 80
    g. 99
    h. 100

2. a. 58
    b. 68
    c. 48
    d. 50
    e. 80
    f. 100
    g. 99
    h. 89

Exit Ticket

a. 99
b. 97

Homework

1. a. 68
    b. 97
    c. 79
    d. 99
    e. 100
    f. 99
    g. 89
    h. 99

2. a. 99
    b. 78
    c. 98
    d. 89
    e. 79
    f. 88
    g. 99
    h. 100

Lesson 13

Problem Set

1. a. 91
    b. 91
    c. 83
    d. 83
    e. 93
    f. 91

2. a. 61
    b. 93
    c. 96
    d. 82
    e. 96
    f. 95

Exit Ticket

a. 86
b. 94

Homework

1. a. 41
    b. 95
    c. 82
    d. 82
    e. 92
    f. 88
    g. 92
    h. 93
    i. 91

2. a. 81
    b. 83
    c. 91
    d. 96
    e. 82
    f. 93
    g. 92
    h. 95
    i. 81

Lesson 14

Problem Set

1.
   a. 69
   b. 70
   c. 82
   d. 82
   e. 91
   f. 94
   g. 95
   h. 97

2.
   a. 70
   b. 81
   c. 100
   d. 95
   e. 93
   f. 82
   g. 98
   h. 96

Exit Ticket

a. 89
b. 100
c. 94

Homework

1.
   a. 89
   b. 91
   c. 83
   d. 94
   e. 93
   f. 94
   g. 95
   h. 87

2.
   a. 80
   b. 91
   c. 100
   d. 84
   e. 92
   f. 94
   g. 88
   h. 96

Lesson 15

Problem Set

1. a. 71
   b. 93
   c. 79
   d. 82
   e. 93
   f. 77

2. a. 71
   b. 100
   c. 82
   d. 82
   e. 82
   f. 95

Exit Ticket

a. 83
b. 93

Homework

1. a. 81
   b. 84
   c. 79
   d. 81
   e. 96
   f. 85

2. a. 84
   b. 90
   c. 92
   d. 92
   e. 82
   f. 96

Lesson 16

Problem Set

1.
   a. 72
   b. 83
   c. 84
   d. 79
   e. 83
   f. 100

2.
   a. 63
   b. 94
   c. 92
   d. 95
   e. 95
   f. 97

Exit Ticket

a. 75
b. 95
c. 92
d. 95

Homework

1.
   a. 84
   b. 92
   c. 85
   d. 80
   e. 86
   f. 98

2.
   a. 93
   b. 75
   c. 91
   d. 85
   e. 98
   f. 97

Lesson 17

Problem Set

1. a. 91
   b. 90
   c. 89
   d. 94
   e. 85
   f. 97

2. a. 71
   b. 79
   c. 92
   d. 95
   e. 100
   f. 93

Exit Ticket

a. 86
b. 90
c. 93
d. 97

Homework

1. a. 82
   b. 100
   c. 79
   d. 94
   e. 95
   f. 97

2. a. 81
   b. 89
   c. 99
   d. 95
   e. 100
   f. 93

Lesson 18

Problem Set

1. 95
2. 100
3. 80
4. 92
5. 49
6. 53

Exit Ticket

Student A work circled; Student B work corrected using number bond under 56 showing 50 and 6; 35 + 50 = 85; 85 + 6 = 91

Homework

1. 76
2. 67
3. 82
4. 73
5. 85
6. 86

Lesson 19

Problem Set

1. 64
2. 84
3. 100
4. 100
5. 89
6. 91
7. 83
8. 81
9. 61
10. 94
11. 68
12. 97

Exit Ticket

1. 62
2. 72

Homework

1. 75
2. 75
3. 90
4. 92
5. 90
6. 100
7. 74
8. 94
9. 74
10. 94
11. 72
12. 94

Lesson 20

Problem Set

1. a. Dime
   b. Penny
   c. Nickel
2. a. 9 pennies drawn
   b. 4 pennies drawn
3. Hand with 2 dimes crossed off
4. 2 nickels drawn; 1 nickel and 5 pennies drawn
5. Not correct; 3 cents is less than 5 cents.

Exit Ticket

1. a. Matched with the nickel
   b. Matched with the dime
2. 5 pennies drawn

Homework

1. Coins appropriately matched
2. a. 2 pennies crossed off
   b. 7 pennies crossed off
3. 1 nickel drawn; 5 pennies drawn
4. a. 20; line drawn to 2 dimes
   b. 5; line drawn to 1 nickel
   c. 10; line drawn to 1 dime
   d. 1; line drawn to 1 penny

Lesson 21

Problem Set

1. a. 25 pennies drawn
    b. Answers will vary.
    c. Answers will vary.
    d. Answers will vary.
    e. 5 nickels drawn
    f. 1 quarter drawn

2. a. Quarters
    b. Pennies
    c. Dimes
    d. Nickels

3. Answers will vary.

4. a. Matched with nickel
    b. Matched with dime
    c. Matched with quarter

Exit Ticket

a. Nickel
b. Penny
c. Quarter
d. Dime

Homework

1. a. Penny
    b. Dime
    c. Quarter
    d. Nickel

2. a. 10
    b. 1
    c. 5
    d. 25

3. Answers will vary.

4. a. Answers will vary.
    b. Answers will vary.

A STORY OF UNITS

Lesson 22 Answer Key 1•6

Lesson 22

Problem Set

1. a. Penny
   b. Dime
   c. Quarter
   d. Nickel
2. a. Matched with dime
   b. Matched with nickel
   c. Matched with quarter

3. Hand with the nickel crossed off
4. Not correct; answers may vary
5. a. 5 cents, matched with nickel
   b. 10 cents, matched with dime
   c. 25 cents, matched with quarter
   d. 1 cent, matched with penny

Exit Ticket

Both sides of coins appropriately matched with name

Homework

1. a. 5, matched with nickels
   b. 10, matched with dimes
   c. 25, matched with quarters
   d. 1, matched with pennies

2. Answers will vary.
3. Answers will vary.

Module 6: Place Value, Comparison, Addition and Subtraction to 100

Lesson 23

Problem Set

1. a. 3 pennies drawn
   b. 5 pennies drawn
   c. 4 pennies drawn
   d. 5 pennies drawn

2. a. 13 cents
   b. 12 cents
   c. 30 cents
   d. 34 cents
   e. 31 cents

Exit Ticket

a. 4 pennies drawn
b. 4 pennies drawn

Homework

1. a. 5 pennies drawn
   b. 3 pennies drawn
   c. 7 pennies drawn
   d. 7 pennies drawn

2. a. 22 cents
   b. 15 cents
   c. 27 cents
   d. 31 cents
   e. 32 cents

Lesson 24

Problem Set

1. a. 3, 2; 30 + 2 = 32
   b. 12, 0; 120 + 0 = 120
   c. 11, 4; 110 + 4 = 114
2. a. 8 dimes checked; 8, 0
   b. 10 dimes checked; 10, 0
3. Five dimes and eight pennies drawn; 5, 8

Exit Ticket

4, 4; 40 + 4 = 44

Homework

1. a. 2, 1; 20 + 1 = 21
   b. 11, 0; 110 + 0 = 110
   c. 11, 3; 110 + 3 = 113
2. 11 dimes checked; 11, 0
3. a. 7 dimes and 9 pennies drawn; 7, 9
   b. 11 dimes and 8 pennies drawn; 11, 8

Lesson 25

Problem Set

1. 10
2. 10
3. 14
4. 13
5. 7
6. 7

Exit Ticket

4

Homework

1. 10
2. 10
3. 14
4. 15
5. 6
6. 8

Lesson 26

Problem Set

1. 6
2. 19
3. 2
4. 12
5. 17
6. 6

Exit Ticket

17

Homework

1. 7
2. 17
3. 6
4. 18
5. 11
6. 18

Lesson 27

Problem Set

1. 4
2. 11
3. 6
4. 7
5. 18
6. 5

Exit Ticket

12

Homework

1. 4
2. 8
3. 7
4. 6
5. 8
6. 13

Lesson 28

Sprint

Side A

1. 2
2. 3
3. 4
4. 3
5. 1
6. 4
7. 5
8. 4
9. 6
10. 7
11. 5
12. 4
13. 6
14. 8
15. 7
16. 9
17. 8
18. 10
19. 7
20. 6
21. 9
22. 10
23. 9
24. 8
25. 10
26. 7
27. 10
28. 9
29. 8
30. 10

Side B

1. 1
2. 2
3. 1
4. 4
5. 3
6. 5
7. 4
8. 5
9. 7
10. 6
11. 8
12. 6
13. 5
14. 7
15. 6
16. 8
17. 9
18. 7
19. 8
20. 10
21. 9
22. 10
23. 10
24. 10
25. 7
26. 8
27. 10
28. 9
29. 9
30. 9

Problem Set

1. Answers will vary.
2. Answers will vary.

Homework

1. All boxes checked
2. 92, 94, 95, 96, 97, 98, 99, 100, 101, 102, 103, 104, 106, 107, 108, 109, 110, 111, 112, 113, 114, 115, 116, 117, 118, 120
3. 87, 67, 57, 47, 37, 27, 17, 7
4. Answers will vary.

Lesson 29

Number Bond Dash

1. 0
2. 1
3. 2
4. 1
5. 0
6. 1
7. 2
8. 3
9. 2
10. 3
11. 4
12. 3
13. 4
14. 5
15. 6
16. 4
17. 6
18. 7
19. 6
20. 7
21. 10
22. 9
23. 8
24. 6
25. 8

Lesson 30

Culminating Activities

Answers will vary.